The Science Race

The Science Race:
Training and Utilization of Scientists and Engineers, US and USSR

Catherine P. Ailes
Francis W. Rushing

Crane Russak • New York

The Science Race:
Training and Utilization of
Scientists and Engineers, US
and USSR

Published in the United States by

Crane, Russak & Company, Inc.
3 East 44th Street
New York, New York 10017

Library of Congress Cataloging in Publication Data

The Science race.
Bibliography: p.
1. Scientists—United States. 2. Engineers—United States.
3. Scientists—Soviet Union. 4. Engineers—Soviet Union.
5. Scientists—Training of—United States. 6. Engineers—Training
of—Soviet Union. I. Ailes, Catherine P. II. Rushing, Francis W.
Q149.U5S313   331.7'9   81-17516
ISBN 0-8448-1407-5   AACR2

Printed in the United States of America

SRI INTERNATIONAL is an independent nonprofit organization providing specialized research services under contract to business, industry, the U.S. government, and some foreign governments, particularly those in the developing nations. Since its foundation in 1946 in Menlo Park, California, the Institute's basic aims have been to enhance economic, political, and social development and to contribute through objective research to the peace and prosperity of mankind.

The Strategic Studies Center of SRI was organized in 1954 by Richard B. Foster, Director. Based in Washington, D.C., the Center conducts multidisciplinary research on the crucial issues of foreign, defense, and international economic policy. With a client structure consisting of the key U.S. government agencies charged with responsibility in these areas, the Strategic Studies Center has long contributed to the ongoing dialogue in both the policymaking and research communities on the critical choices facing the United States, particularly in the field of national security.

To make the timely findings of the Center's research available to a broader public, the Strategic Studies Center is publishing a series of books and monographs, of which this volume on Africa is another in the series of books. The views expressed are those of the author and do not necessarily reflect the position of the Center.

Preface

This book reports on an extensive comparative study of the training and utilization of scientists and engineers in the United States and the Soviet Union. The research was an outgrowth of years of exchanges of information, published reports, and other scholarly interaction between U.S. and Soviet members of the Subgroup on Training and Utilization of Scientists and Engineers of the Science Policy Working Group under the U.S.-USSR Scientific and Technical Exchange Program. This book has effectively synthesized the information exchanged by this group. The exchange materials have been liberally supplemented and expanded by the authors to round out the picture of how the contrasting U.S. and Soviet systems train and utilize scientists and engineers to achieve their respective national objectives.

As a consequence of this comparative analysis, a number of issues have surfaced which are important to U.S. government policy makers, members of academia, and the business and commercial leadership. The most dramatic contrast is the difference in the national commitments of the United States and USSR to prepare scientific and engineering manpower to meet the technological challenges of the future.

Although this study is not an official publication of the U.S. side of the working group and only the authors are responsible for its content, I wish to take this opportunity to thank all those who assisted in its preparation.

William D. Carey
U. S. Chairman, U.S.-U.S.S.R.
Working Group on Science Policy &
Executive Officer of the American
Association for the Advancement
of Science

August 1981

Authors' Preface

This book presents a comparison of the systems of training and utilization of scientists and engineers in the United States and the Soviet Union. Chapter I provides a general description of the economic structure and organization in which the training of scientists and engineers is conducted and in which such trained personnel are employed. In Chapters II through V, the systems of training scientific, engineering, and technical personnel in the two countries are described. The discussion covers the structure of the two educational systems, focusing on general education, training of technicians, higher education, and graduate training. Statistical tables providing a quantitative comparison of entrance, enrollment, and completion of the various stages in the educational process in the two countries are included where appropriate. An attempt has been made when possible to normalize the data that are presented in terms of relevant age group populations, labor force, or GNP.

Chapters VI through IX of this study focus on the employment aspects of scientists and engineers in the two countries. The discussion covers the definitions used for classifying individuals as scientists, engineers, or technicians in the two countries, the utilization of scientists and engineers in R&D employment sectors, and the mobility of scientists and engineers between fields, positions, and geographical areas. Statistical tables showing trends in employment in science and engineering in the Soviet Union and in the United States are presented, as well as U.S.-Soviet comparisons of total scientific and engineering manpower and estimated R&D employment aggregates. The principal methods used in the two countries to forecast the demand for and supply of scientists and engineers are described, and some projections of the supply of scientists and engineers are presented to give the reader some indicators of future comparative growth rates. Finally, Chapter X summarizes and assesses some of the principal findings of this study.

The comparisons presented in this book were prepared at the request of the U.S. members of the Joint Subgroup on Training and Utilization of the Science Policy Working Group under the U.S.-U.S.S.R. Science and Technology Exchange Program. While not part of the documents officially exchanged by the subgroup, the book draws heavily on the documents that were formally exchanged by the United States and the USSR under this program of cooperative activity. A complete list is shown in the beginning section of the Bibliography.

As the Soviet and U.S. basic reports and other materials exchanged by the Joint Subgroup serve as the principal source for the descriptive material contained in this study, specific page references to the texts of those materials are not included. Where data contained in any of those materials are used, we have referred to the primary source of the statistical information, except where otherwise indicated.

There are a number of individuals who deserve a special note of appreciation for their support and contributions to this research. Our greatest debt is owed to Richard B. Foster, Director of the Strategic Studies Center, who provided invaluable insights into Soviet scientific and technological objectives and who has been a constant source of support and encouragement for the project since its inception. We also want to express our appreciation to James G. Styles, who provided valuable research assistance in the preparation of this study.

Portions of the chapters on education were submitted to the National Science Foundation as an input paper for the White House Report, "Science and Engineering Education for the 1980s and Beyond." We again wish to express our appreciation to Nicholas DeWitt of Indiana University for his very helpful comments and suggestions on those portions of this study.

We have greatly benefited from the continued support of Joel L. Barries of the National Science Foundation. Credit must also be given to Dael Wolfle, who served as head of the U.S. side of the Subgroup on Training and Utilization; Murray Feshbach, Simon Kassel, Charles V. Kidd, and Thomas J. Mills, who served as members of the U.S. side of the Subgroup, and Robert W. Campbell, Herbert S. Levine, and Louvan E. Nolting, who participated in the Subgroup during many of its discussions. Finally, we wish to express our gratitude to the Soviet and American scholars who contributed to the various reports and documents noted in the Bibliography.

Foreword

When the Soviet Union launched Sputnik in 1957, the specter of Soviet scientific supremacy shocked the United States. Americans, long accustomed to being the most technologically advanced people in the world, suddenly were confronted with the possibility that their military rival had overtaken them. Stunned by the implications of this unexpected development, the U.S. government massively increased funds for education and research in the sciences, viewing these expenditures as necessary for the safeguarding of American national security.

After a short while, however, it became clear that American fears of technological inferiority were unfounded—at least in the short run. The United States, challenged by the Soviet Union to a space race, overcame its initial difficulties and became the first—and still the only—country to land men on the moon. Their decisive triumph in the competition in space reassured Americans that they were indeed technologically superior to the Soviet Union, and the previous concern about relative scientific capabilities waned.

Yet, even while Americans have become complacent, the technological balance has been shifting against them. By conscientious effort over the years, the Soviet Union has increased its efforts in the training of scientific and engineering personnel to the point that its programs now dwarf those in the United States. Because this effort has proceeded silently, it has gone largely unnoticed in the United States. If the United States fails to respond, however, there can be little doubt that eventually it will find itself technologically inferior to the Soviet Union.

A key element in Soviet economic, political, and military policies in response to increasing labor constraints and limitations on natural resources is the application of science and technology to production to bring about an intensive transformation of the economy. Such an approach, if successful, could have profound effects on Soviet economic performance, and thereby broaden the economic foundation necessary to support Soviet military and political objectives without sacrifice in terms of other national objectives.

Soviet statements, as well as the dramatic growth in the funding of research and development and the training and employment of scientists and engineers, suggest a strong drive to catch up with Western technology and a firm commitment to transform the technological level of the economy in order to improve economic growth prospects. Technological modernization of the economy has been a theme of Russian/Soviet policies since the nineteenth century, and in fact can be traced back to the reign of Peter the Great at the turn of the eighteenth century, but it appears to have escalated greatly in the last twenty years.

The growth of cadres of scientists and engineers in the USSR has indeed been dramatic, as this comparative study makes clear. Although the research does not specifically separate out the military sector, the study's findings do suggest some hypotheses that warrant U.S. national concern and require rigorous testing and analysis by further research.

The Soviet military establishment is indeed closing the qualitative gap with the West, and the United States in particular. The United States can no longer count on superior technology to offset the Soviet numbers of men and masses of weapons. There are still examples of U.S. superiority, as in the computer and electronic component-miniaturization fields, but there are also increasing numbers of areas in which the Soviet Union appears to be catching up or passing the United States, as in missile accuracy, tank armor and armaments, antitank guided missiles, infantry fighting vehicles, and naval antiship missiles (including nuclear). It is time for a new, up-to-date net evaluation of these technological differences.

Despite the earlier noted priority of military claims on technological brains and equipment and past failures in modernizing important civilian sectors (notably, agriculture), the Soviet Union in the future may well allocate sufficient technology to the civilian economy to increase productivity and forestall the economic difficulties foreseen in the more dire Western forecasts. At the present time, the Soviet Union imports many high-technology items, buying both finished products (e.g., oil drill bits) and turnkey plants. They have great incentives to train people qualified to utilize the technology in their own products and plants, so as to avoid continuing importation.

Soviet training levels in science and engineering, and especially the much higher mathematical training in the primary and secondary grades, mean an increasingly qualified manpower pool for the military, in contrast to the reportedly rapidly declining quality of volunteers in U.S. forces, particularly the U.S. Army. It may be that the growing complexity of military equipment finds Soviet enlisted men and noncommissioned officers increasingly qualified to operate and maintain the forces—and U.S. personnel decreasingly so. The United States may have to change both its educational and its military manpower strategies, and perhaps also its weapons design philosophy.

The importance of these issues to Soviet leadership objectives and thus to the challenge to be faced by Western nations prompts one to consider carefully hypotheses that may be less comforting and less conservative in terms of the willingness and ability of the Soviet leadership to institute a basic transformation of the economy when it is perceived as a categorical imperative. Such an imperative was recognized in the sphere of military capabilities early in the 1960s and resulted in the surprise of quantitative and qualitative strategic parity with the West in the early 1970s. This should not have been a surprise, given Soviet doctrine and the commitment to the effort evidenced by large allocations of resources.

Similar doctrinal and resource trends are also in evidence with respect to the Soviet science sector, in terms of both the financing of research and development activity and the training of scientific and engineering manpower. The size of the commitment of resources to these efforts, and the remarkable rates of growth, cannot be dismissed simply as functions of the low base from which they proceed and inefficiency in utilization of resources, or even as just an all-out effort to catch up somewhat with the level of Western technology; such a complacent attitude reflects on underestimates of the wealth of evidence provided in the Soviet literature that this is the fundamental approach to future development and the clear perception by Soviet policy makers that there is no other course to follow without a serious narrowing of long-term objectives. The response of the Soviet leadership to the Cuban missile crisis of 1962 was not to narrow objectives, but to increase the commitment to a long-term program that would improve their strategic options.

The priority of military efforts throughout the economic system was responsible for much of the achievement over the earlier period, yet as priorities multiply they lose their meaning. The application of scientific and technical achievements throughout the economy is thus a much more ambitious task and requires basic organizational change. Since that change has fallen far short of what is required, recent trends have remained unfavorable. Failure to meet the basic resource requirements of key military and civilian programs, however, may well induce change that is evolutionary but far from counter-revolutionary, and the transformation of economic constraints may well ensue. If the policy were successful, the Soviet Union could reduce the impact of current and projected economic constraints, providing an economic foundation for accelerated programs in the military arena without increased sacrifices in terms of other national objectives, and become an ever more formidable adversary in the future.

The Soviet Union is firmly committed to the achievement of scientific and technological superiority not only as a major goal in itself but also as a prerequisite for the attainment of other strengths and capabilities to increase the chance of Soviet success in its competition with the West and the victory of the socialist over the capitalist system. In 1957, with the Soviet launching of Sputnik, the United States rose to the challenge and mobilized its resources for better science and engineering education for America's schoolchildren, increased spending in R&D, and placed greater emphasis on scientific and technological excellence in all areas, viewing these efforts as necessary for U.S. national security. In the past decade, however, Americans have once more become complacent. Recent events amount cumulatively to a new Sputnik, but the United States has as yet to recognize the urgency of the problem. Just as Sputnik made the challenge of the space race clear, the consistent pattern of large Soviet gains in scientific and other technical personnel renews the specter of Soviet scientific supremacy and suggests a similar challenge—the science race.

RICHARD B. FOSTER
Senior Director
Strategic Studies Center
SRI International

Table of Contents

LIST OF ILLUSTRATIONS

LIST OF TABLES

CHAPTER I

Economic Allocation in the Soviet Union and the United States

It is useful to begin a comparison of the training and utilization of scientists and engineers in the United States and the Soviet Union with a general description of the economic structure and organization in which such training is conducted and in which such trained personnel are employed. The contrast of the two economic systems may be best described as one being market-oriented and the other being plan-directed. The Soviet plan-directed economy will be described first and then compared with U.S. economic market institutions.

A. The Soviet Plan-Directed Economy

The economy of the Soviet Union is generally described as a plan-directed or command economy. The basic choices that direct the economy are made by the Politburo of the Communist Party and transmitted through various channels to the Council of Ministers of the Soviet Union, which then formulates expanded definitions of the goals set by the Politburo. The State Planning Committee (Gosplan) has the responsibility of designing a comprehensive plan for which specific directives are issued and compliance is required.

Historically, planning has been for two time periods: the five-year plan with specific identified objectives, and the annual plan specifying incremental steps toward achieving the five-year-plan goals. The annual plan has historically been law, while the five-year plan, except for the most recent one (1976-80), has not been. In the 1970s, the Soviet leadership initiated longer-term economic planning, such as the fifteen-year plan which is believed by most Western observers to exist but has not been published for general distribution. Current science policy is also expressed in long-term plans (from 5 to 20 years) for science and technology. The operative economic plan with respect to management decisions is the annual plan. However, the longer-term plans covering five or more years are particularly important in the discussion of training and utilization of scientists and engineers in the USSR. Their importance is derived from the fact that future skilled manpower requirements have a four to six year lead-time to fill.

The Soviet economic growth model has traditionally relied heavily on the allocation of additional inputs in order to produce additional outputs. Even within the context of this extensive growth model, however, the Soviet leadership has placed heavy emphasis (and resources) on the education of the population to raise the skill level of the work force, with the intended outcome being greater productivity. The design of the educational system reflects the demographic realities of the USSR and the fact that the system considers each student as a future input to the economic system. Some form of education and/or training is available to Soviet citizens during all periods of their lives. Universal education, grades 1–8, is only the earliest educational stage. As Chapter II will discuss, there are several educational options after grade 8. Even a full-time employed individual has opportunities, and many times strong encouragement, to continue his education. The many and varied educational and training opportunities in the Soviet Union are consistent with announced Soviet objectives of increasing output through a technological transformation of the economy. As technologies are changing rapidly, training of workers is essential for the full utilization of economic resources.

The Soviet planners' objective is to train and utilize manpower in the most effective way possible so as to achieve the state's political and economic goals. Within the parameters determined by the state goals and facilitated by the state planners, the individual attempts to maximize his or her own well-being. While the individual within this economic system is free to follow any of the various channels open to the future, those channels are directed and redirected according to the dictates of the plan.

Figure I-1 provides a schema of the Soviet planning process over a period of time. The time dimension permits the planners to alter the number and quality of labor inputs. Given the party directives, the Soviet planners first must specify broad economic goals and objectives. These objectives can normally be specified as unique outcomes stated as civilian and military goods and services. The planners are expected to maximize these outputs given the constraints imposed upon them by technology, quantity and quality of labor, and quantity and quality of capital. Given enough time, each of these variables can be changed.

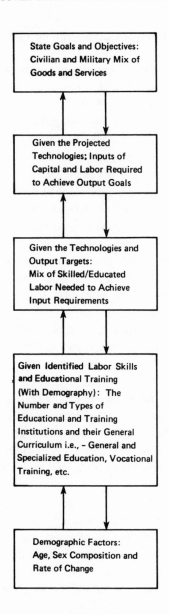

State Goals and Objectives:
Civilian and Military Mix of
Goods and Services

Given the Projected
Technologies: Inputs of
Capital and Labor Required
to Achieve Output Goals

Given the Technologies and
Output Targets:
Mix of Skilled/Educated
Labor Needed to Achieve
Input Requirements

Given Identified Labor Skills
and Educational Training
(With Demography): The
Number and Types of
Educational and Training
Institutions and their General
Curriculum i.e., – General and
Specialized Education, Vocational
Training, etc.

Demographic Factors:
Age, Sex Composition and
Rate of Change

Denotes
Information
Flow

Figure I-1   SOVIET PLANNING OF MANPOWER FOR MEETING STATE GOALS

The quantity of labor (a demographic factor) is the most difficult to alter by state action.

Future growth and composition of outputs can be achieved by known technologies and by projected new and improved technologies. Under current Soviet policy, the direction and pace of technological change are variables which are to be manipulated by effective planning. Once the outputs and the technologies necessary to achieve the outputs are specified, the inputs of capital and labor, both quality and quantity, can then be determined. Because of the focus of this study, only the labor component of these inputs will be discussed.

Soviet economic planners have traditionally expanded the quantity of labor inputs in the industrialized sector by transferring labor out of agriculture and by increasing the labor participation rates of the population, particularly the female component, although retirement age has been adjusted to retain persons in the labor force for longer periods of time. In addition, the educational institutions have been modified periodically to increase the supply of labor. This was accomplished by shortening the formal educational period, thus permitting youths to enter the labor force earlier. In addition, the government encouraged young people, except the most outstanding students, to enter the work force and then continue their education on an evening or part-time basis. Finally, the labor supply can be reallocated (although not actually expanded in number) by either redirection from lower to higher priority sectors or by labor being released through enhanced productivity due to better labor incentives and/or technological change. This latter alternative is currently being promoted as the best solution to the quantity of labor constraint.

The quality of labor is primarily a function of the education and training of Soviet citizens. Given the future output mix and the anticipated technologies, Soviet planners must then ensure that the educational and training institutions are sufficient in number and have appropriate curricula to meet the future input requirements. The composition of fields of training and the number of persons trained in each field reflect the overall plan. Thus, as Soviet citizens progress up the educational ladder, the options open to them are determined not only by their past performance and future potential but also by the preferences of the state which are manifested in the number of positions in each field which open up annually. The specific number of openings in each field of training reflects the planners' judgment of the number of employment positions which will be available when the education and training period is completed.

Soviet workers, like their U.S. counterparts, are influenced by the anticipated economic and social/political rewards from pursuing certain career paths. First and foremost, it seems that education is the best path of upward mobility within the Soviet system and careers within some specific educational areas have more promise than others—particularly engineering and scientific occupations. While earnings and other benefits influence the choices of individuals, the state planners define the parameters and therefore the numbers of persons who can successfully pursue their individual objectives in any particular field.

B. The U.S. Market-Oriented Economy

The major contrast between the market-oriented system and the plan-directed system is in terms of the extent to which individuals have the freedom to exercise their choices. In the Soviet system, the state plans dominate both supply and demand, while in the U.S. case, the decisions of individuals, when aggregated, help determine the supply side of the market, while the demand for labor represents a summation of business, government, and academic needs for scientists and engineers.

The chain of events involved in the training and employment of individuals in scientific, engineering, and technical fields in the United States is dependent on the existence and operations of markets. A market is defined as the interaction of independent suppliers and demanders which has as an outcome the exchange of goods and services at prices which the participants are willing to accept. Markets are also transmitters of information—information which forms the basis upon which the suppliers and demanders take action.

In the U.S. economy, individuals are permitted to make educational and career decisions for themselves, subject to the constraints that various markets place upon them and, in the early years, to government requirements to remain in the educational system until 16 years of age. Given this system of markets, a general theory of individual economic behavior has been developed to describe the decision-making process by which individuals select their educational paths and their careers.

Figure I-2 provides a schema of the determinants of the supply and demand for scientists and engineers in the U.S. economy. The basic characteristics of the model are that the supply of scientists and engineers is a function of demographic factors and a series of individual decisions with respect to training and career choices—most of which are economic-related and subject to market-related influences.

The demand for scientists and engineers, as a function of the current and expected needs for trained persons, reflects three major demand segments—business, government, and academe. Each of these demand

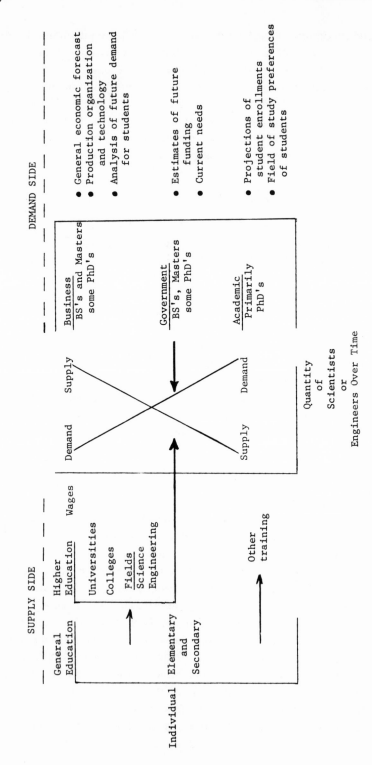

Figure I-2 MARKET SUPPLY AND DEMAND FOR SCIENTISTS AND ENGINEERS
IN THE UNITED STATES

components estimates its requirements based on factors such as projected business activities, nature and level of government-funded programs, college enrollments, etc.

In the case of business, the demand for scientists and engineers is based on the demand for the product as well as the production organization and technology of the particular business or industry. Production technology helps determine the most efficient combination of labor and capital and the general productivity of each factor. It also helps determine the quantity of labor of various skills needed. Finally, the general determinant of wages and salaries is the contribution of labor to the value of output—thus the more productive a worker, the higher his or her salary. This relationship between productivity and wages and salaries implies that the better trained the worker, either through experience, education or both, the higher his income.

For analytical purposes, economists generally assume that individuals seek to maximize their incomes and therefore choose the kinds of education and training which will lead to the highest income possible for them. There are, of course, other important individual considerations in career choice: innate abilities, life-style preference, proximity to relatives, geographic areas, rural-urban choices, etc., all seem to play some role in career and job choice and are sometimes listed ahead of income in surveys conducted as part of sociological or psychological studies. However, income has received the most attention by economists, in part because it can be used effectively in predictive models.

Economists describe investment in education or training as investment in human capital. Individuals, to achieve their objectives, must constantly look at market indicators to determine the best opportunities to maximize their income given their individual constraints—i.e., natural ability, financial limitations, education, etc. Given his level of training, the individual will then elect to accept an employment position which maximizes his well-being. The individual, unfortunately, is limited to those choices available at a point in time. Opportunities may be different from those anticipated when the individual entered the education and training process.

In the United States, it is believed that the market will attract or discourage prospective entrants by raising or lowering salaries or by the availability of jobs within certain fields. Generally, market information is effectively transmitted, but there are significant time lapses before supply adjustments are completed due to the fact that training takes time. In addition, market signals are sometimes misleading because of new technologies being introduced that change the demand for certain types of specialists.

A distinction should be made between private returns and social returns to human capital investment in the market economy. Private returns are those which accrue to the individual undertaking the investment. Social returns are those which accrue to society, and consist of both the individual's return and the net effects of the investment in that individual on the rest of society.

Private returns to human capital investment are the difference between the personal costs to the individual undertaking the investment and increases in his personal income. An individual has a range of possible returns, determined by his ability and by the duration and type of available human capital investment opportunities. Economic rationality dictates that the optimal level of investment for an individual is that which maximizes his expected rate of return adjusted for risk. However, the typical pattern of income associated with any investment begins with a period of low returns during training time followed by a period of higher returns from increased wages.

There are several issues pertaining to private returns on human capital investment. First, the investment decision process is continual. An individual can change his human capital investment at any time—that is, he can respond to changes in the labor market by changing the quantity and/or type of his education and training.

Second, foregone earnings are the largest cost component of human capital investment. Time spent on education or training is time spent away from productive (and therefore income-generating) activity. To the extent that an individual works part time while in formal learning, his costs in terms of foregone earnings are reduced or are spread over more years.

Third, increasing age reduces the private rate of return on human capital investment, ceteris paribus, because an older individual has fewer years in which to receive benefits from his investment. As a result, older individuals are less likely to undertake human capital investment.

Social returns to human capital investment are determined by the difference between the total cost to society of the resources required for the investment in the individual and the total benefit to society (including "externalities"[1]) which results from this investment. The social cost of human capital includes all government and private-in-

---

[1] In economics, externalities are any costs or benefits which accrue to neither the producer nor the consumer of a particular good or service, but rather to a third party who is outside of the market transaction in question.

dustry financial support for education as well as the direct cost to the individual for increasing his stock of human capital.

It is not currently possible to estimate accurately social rates of return; it is clear, however, that in many cases social returns differ substantially from private returns. For example, one person's human capital investment could result in the discovery of a new technology that raises the productivity of others. The private return to the human capital investor would be only his personal income increase and would not include all of the income increase of others whom he helped to make more productive. The social return, however, consists of the total production increase for society.

Private decisions in the United States are based upon private costs and benefits, and therefore do not necessarily represent the socially optimal level of human capital investment. Theoretically, the government could induce a socially optimal level of human capital investment. The U.S. government, of course, does not regulate the labor market or scientists and engineers. Government-funded programs, however, work to increase the private gains to individuals in certain selected fields, and thus, while allowing individuals to maintain occupational freedom of choice, simultaneously encourage them to select fields which the government wishes to emphasize.

C. Summary

The contrast between plan and market mechanisms for directing the training and utilization of scientists and engineers is very dramatic. However, a direct assessment as to which is the better of the two systems is difficult, as the criteria for evaluation of either system must include how effectively the methods utilized by the respective systems help achieve the goals of that society.

In terms of the goals of the Soviet Union specified by the Communist Party, the plan-directed approach to resource allocation is both feasible and, in a number of ways, also successful. The Soviet economy has grown and expanded and has facilitated the rise of the USSR to superpower status—both militarily and economically. The system has also permitted the continued control by the party of not only the economy but the entire society. The primary question now being posed by the Soviet leadership, as well as by Western observers, is whether the planning system, as it has existed in the last 52 years, can meet the new requirement of the economy of changing from an extensive to an intensive growth process. There are, however, no clear signs of any major reorganization of the Soviet economy now under way.

The market economy is guided by the choices expressed by individuals and their representative government. Markets have generally received high marks for efficiency of allocation, although the nature of the market sometimes fails to fully take into consideration the national or societal interest.

In the U.S. economy, individual decisions based on market conditions are the principal determinants of work-force training and utilization. However, there is still significant government influence on these decisions. The government, at the local, state, and federal level, provides public education from general tax revenues through the secondary education program. State and federal government funds heavily subsidize public institutions of higher education. Federal funds, in many cases, provide significant financial resources to private institutions of higher education. In addition, the activities of government in the marketplace impact on the kinds and numbers of persons needed to provide the goods and services for the U.S. economy—for instance, the defense-related industry. Thus, although the focus is on the individual and the market, the government plays an important role in economic decisions in the U.S. economy.

In the Soviet Union, government policies are designed to maximize the social returns to human capital investment, with private returns being of secondary importance. The absence of certain property rights for individuals, such as rights to own means of production, patents, product design, etc., redirects what would otherwise be private returns to the state.

In the United States, the direct returns to human capital investment generally accrue to the individual, but social returns are more difficult to capture in a market-oriented economy. Thus, the U.S. worker may enjoy the fruits of his training and also bear the burden of his miscalculation based on market information. The state in the Soviet Union tends to reduce the risk to the individual by permitting choices from among fields which the state desires to have expanded or continue at the existing levels of employment. The trade-off for the individual is in terms of limitations on his or her choices. In the Soviet Union, however, in contrast to the United States, the state, rather than the individual, bears the major financial responsibility for the education or training in the post-secondary years. For part-time students there are generally periods in which the employee/student is permitted time off with pay to complete his educational or training requirements. While this is sometimes also true with respect to U.S. workers, it is not on the same scale as in the USSR.

The two systems, although different in structure of decision-making and resource allocation, also have similarities. The United States and the Soviet Union are both large industrial economies with a continuing and enhanced need for high-quality, skilled labor and well-trained scientists and engineers. In the chapters that follow, both the differences and similarities in the training and utilization of scientists and engineers will be described.

CHAPTER II

The Educational Systems:
An Overview

The contrast between the U.S. and Soviet educational structures clearly reflects the differences in the goals and objectives of education within the two political-economic systems. The Soviet educational structure is essentially designed to provide educated persons to the labor pool for the national economy. To meet those requirements, the USSR has established a national structure of education, with standardized curricula and common examination procedures.

This contrasts directly with the U.S. system, which is neither centralized nor standardized (although there are general characteristics which tend to permeate the system) and where the goals of the individual are of equal or greater importance than those of the state. While one of the most prominent influences on the structure of education is that of educating individuals who can enter into the world of work, there are still very important considerations given to designing curricula which also contribute to the development of the individual. Thus, in the United States, courses in sociology, art, history, etc. are believed to be important in helping the student to learn to interact with other people in a society where freedom of individual choice is broader than in Soviet society.

Both the United States and the Soviet Union are committed to education and have devoted a significant portion of their national resources to its development. Table II-1 shows U.S. and USSR total expenditures on education from 1960 to 1979. As a percent of GNP, Soviet expenditures on education remained at about 5.5 percent from 1966 through 1977, the years for which GNP estimates are available. In the United States, there was a substantial increase in total expenditures on education as a percent of GNP from 1960 to 1975, rising from 5.0 percent in 1960 to 7.6 percent in 1975. In the latter part of the 1970s, U.S. expenditures on education as a percent of GNP have

TABLE II-1

U.S. AND U.S.S.R. TOTAL EXPENDITURES
ON EDUCATION: 1960-1979

	1960	1966	1970	1975	1976	1977	1978	1979
U.S. Total (Billions of Dollars)	24.7	45.2	70.4	111.1	121.8	131.0	104.4	151.5
(Percent of GNP)	5.0	6.3	7.3	7.6	7.5	7.3	7.1	6.7
U.S.S.R. Total (Billions of Rubles)	8.5	14.1	19.8	26.2	27.2	28.2	29.5	30.4
(Percent of GNP)	NA	5.4	5.5	5.8	5.4	5.3	NA	NA

[1]   Including expenditures from the State Budget (which account for 75 to 80 percent
        of total expenditures on education) and expenditures financed by collective farms,
        trade unions and other non-budgetary sources.

Sources: Statistical Abstract of the United States, 1980, p. 140;
          National Economy of the USSR in 1979, p. 555;
          National Economy of the USSR in 1969, p. 771;
          Science Indicators 1978, National Science Board (1979), p. 141.

been declining, falling to 6.7 percent in 1979. Although the United
States still devotes a greater share of GNP to education than does the
Soviet Union, one should note that the economy is between one quarter
and one-half larger than the Soviet economy.

The clearly enunciated approach of the Soviet leadership for meeting
the long-term future is based on the application of the achievements
of science and technology in all areas of development. An integral part
of this doctrine is the restructuring of the skill levels of the labor
force—a qualitative approach which is intended to provide the most
effective avenue to quantitative improvements in the economic sphere.
The Soviet Union has made remarkable strides toward its goal of
raising the quality of its work force to bring about the scientific and
technological transformation of the economy. As indicated in Table
II-2, the level of educational attainment of the population, measured
by number of years of training completed, has risen from 5.9 years
in 1960 to 8.7 years in 1977, and is projected[1] to rise to 9.9 by 1985.

[1]USSR: Trends and Prospects in Educational Attainment, 1959–85, National Foreign Assessment
Center, ER79-10344, p. 7 (June 1979).

TABLE II-2

U.S. AND U.S.S.R. EDUCATIONAL ATTAINMENT OF THE POPULATION: 1960-1977

	1960	1970	1975	1976	1977
U.S.[1]					
			Percent		
College, 4 Years or More	7.7	11.0	13.9	14.7	15.4
High School, 4 Years or More	33.4	44.2	48.6	49.4	49.5
Fifth Grade, or More	50.6	39.5	33.3	32.0	31.4
Less than 5 Years of School	8.3	5.3	4.2	3.9	3.7
			Number of Years		
Median Level	10.6	12.2	12.3	12.4	12.4
U.S.S.R.[2]					
			Percent		
Higher	2.8	5.0	6.2	6.4	6.7
Incomplete Higher	1.2	1.6	1.5	1.6	1.5
Specialized Secondary	5.7	8.0	9.8	10.1	10.4
General Secondary	7.4	14.0	17.6	18.6	19.8
Incomplete Secondary	23.3	26.8	27.6	27.4	27.1
Primary and Less	59.6	44.6	37.3	35.9	34.5
			Number of Years		
Median Level	5.8	7.6	8.3	8.5	8.7

1 Data for persons 25 years of age or older.
2 Data for persons 16 years of age or older.

Sources: Statistical Abstract of the United States, 1978, p. 143;
USSR: Trends and Prospects in Educational Attainment, 1959-85,
National Foreign Assessment Center, ER79-10344, p.7, 23 (June 1979).

When these data are compared to data for the United States, there is still a significant disparity, especially in terms of percentage of population who have completed a higher education (15.4 percent of total population in the United States with four or more years of college as opposed to 6.7 percent in the Soviet Union having completed approximately 5 years of higher education). The distribution of the Soviet population at the higher levels of education, however, is far more heavily skewed toward the scientific and technical fields than is that of the United States. This emphasis can be expected to continue into the future.

Difference in educational attainment by years, however, is not a sufficient indicator of the educational level of those persons trained. The structure of education, and the curricula within that structure, are also important. These will be discussed in the remainder of this chapter.

A. The Structure of the Educational Systems in the Soviet Union and the United States

The structure of the Soviet educational process reflects the regime's response to changing Soviet demography and changing requirements of the Soviet economy. Economic and demographic problems have forced the USSR to base its future economic well-being on rapid technological change and higher efficiency. Thus, Soviet general education is designed to channel students into the work force at an early age with a background in science and mathematics sufficient to permit them to function productively in a changing high-technology economy. The Soviet leadership makes available to its citizens many opportunities for education and training during both their pre-employment and post-employment years, and attempts to ensure that the best students are admitted to higher educational institutions and eventually to graduate training.

Figure II-1 depicts the general structures of the U.S. and Soviet educational systems. Both systems start formalized education at the age of 6–7 and finish higher education by age 21–22. Each is based on the concept of compulsory elementary and secondary education to ages 15–16 and each has a high degree of participation in the educational process. The U.S. system has a more general hierarchical structure, with a broader curriculum in which specialization does not take place until the later years of higher education. In the Soviet system, on the other hand, specialization begins to occur at the specialized secondary school and in the extensive network of vocational-technical

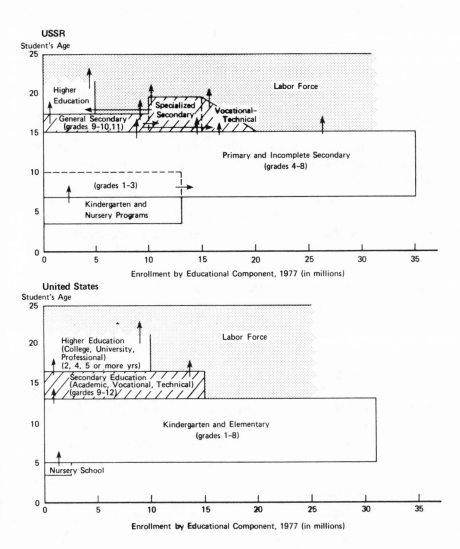

Source: USSR: Trends and Prospects in Educational Attainment 1959–85,
         p. 3, CIA ER79–10344, (June 1979).

Figure II-1 THE STRUCTURE OF EDUCATION IN THE SOVIET UNION AND THE UNITED
             STATES

schools, or at about 16 years of age. This specialization continues into higher education.

B. General Education—Elementary and Secondary

The most dramatic difference between the educational systems of the Soviet Union and the United States is with regard to their curricula—particularly the general educational curricula at the elementary and secondary level. As the curriculum at these grades is the basis upon which subsequent curricula are built, it should be reviewed in some detail.

Figure II-2 summarizes the Soviet curriculum for grades 1–10, breaking it down as the Soviet schools do into primary (grades 1–4), incomplete secondary (grades 5–8), and general secondary (grades 9–10)[2] education. Perhaps the greatest contrast between U.S. and Soviet curricula at this level is the Soviet emphasis on science and mathematics and the grades at which they are introduced. Mathematics is introduced at grade 1, biology at grade 5, physics at grade 6, and chemistry at grade 7. At the terminal point in the ''incomplete secondary training'' (grades 5–8), all students have already had eight years of exposure to mathematics, three years to physics, and two years to chemistry. In addition, the Soviet student has had five years of foreign language training. Those students electing to enter the general secondary curriculum (grades 9–10) will continue the mathematics/science-oriented program. Those students entering specialized secondary institutions will focus on those mathematics and science disciplines that are most appropriate for their field of specialization.

The average number of instructional hours per week devoted to mathematics, science, and social studies in the elementary school program in the United States[3] and the USSR are shown below:

	US	USSR	US	USSR
Grade Range	K-3	1-3	4-6	4-6
Mathematics	3.2	6.0	3.7	6.0
Science	1.6	1.0	2.9	2.3
Social Studies	1.8	–	3.3	1.7

[2]In some Republics where instruction is not in Russian, an eleventh grade may be added to ensure that students are proficient in Russian upon completion of their general education.
[3]See ''The Status of Pre-College Science, Mathematics and Social Studies Educational Practices in U.S. Schools,'' National Science Foundation, NSF-SR-78-71.

Subjects / Grades	Primary School				Eight-Year School				10-Year Complete Secondary School		Total Weekly Hours[1]
	I	II	III	IV	V	VI	VII	VIII	IX	X	
Russian language	12	11	10	6	6	4	3	2	–	–	54
Literature	–	–	–	2	2	2	2	3	4	3	18
Mathematics	6	6	6	6	6	6	6	6	5	5/4	57.5
History	–	–	–	2	2	2	2	3	4	3	18
Basics of Soviet Government & Law	–	–	–	–	–	–	–	1	–	–	1
Social Science		–	–	–	–	–	–	–	–	2	2
Natural Science	–	1	2	1	–	–	–	–	–	–	4
Geography	–	–	–	–	2	3	2	2	2	–	11
Biology	–	–	–	–	2	2	2	2	1	2	11
Physics	–	–	–	–	–	2	2	3	4	4/5	15.5
Astronomy	–	–	–	–	–	–	–	–	–	1	1
Drawing	–	–	–	–	–	–	1	1	–	–	2
Foreign language	–	–	–	4	3	2	2	1	1	1	14
Chemistry	–	–	–	–	–	–	2	2	3	3	10
Fine Arts	1	1	1	1	1	1	–	–	–	–	6
Singing and Music	1	1	1	1	1	1	1	–	–	–	7
Physical Training	2	2	2	2	2	2	2	2	2	2	20
Manual Training	2	2	2	2	2	2	2	2	4	4	24
Primary military training	–	–	–	–	–	–	–	–	2	2	4
Total required courses	24	24	24	27	29	29	29	30	32	32	
Elective courses							2	3	4	4	13

[1] Indicates "instructional hours," which are 45 minutes in duration.

Source: Nicholas DeWitt, "Current Status and Determinants of Science Education in Soviet Secondary Schools," prepared for the National Academy of Sciences, Washington, D.C., p. 14 (April 1980).

Figure II-2 TYPICAL USSR GENERAL EDUCATION SCHOOL CURRICULUM (GRADES 1-10), PROPOSED FOR 1980/81

U.S. elementary school students receive slightly more hours per week in science training than do Soviet elementary students. The accelerated training in mathematics in the Soviet Union at the elementary grades, however, is in sharp contrast to the time spent on these subjects in the U.S. elementary schools. Although the emphasis on mathematics in U.S. elementary schools increases in grades 4–6 as compared with grades 1–3, total hours devoted to mathematics are still considerably below the average for the Soviet Union during the same grades.

The Soviet mathematics curriculum provides the student who terminates his training at grade 8 with two years of algebra and plane geometry. If the student continues through grades 9 and 10, he will have completed two additional years of algebra and two years of calculus. This curriculum applies to *all* students in these grades. Findings from a sample survey in the United States show that over 56 percent of the school systems require no mathematics course or only one for graduation from the secondary school program. Changing patterns for courses in secondary schools in the United States do seem to reveal, however, more and better mathematics training for *some* students, especially advanced college-bound students; these students generally complete a calculus course and perhaps a course in probability and statistics. For most students, the general level of education in the United States seems to provide at least some training in general mathematics, geometry, and algebra (basic and advanced). A recent study of science education found, however, "that one-half of all high school graduates take no mathematics or science beyond the 10th grade and only one-half of the students entering college have had any significant exposure to physical science or advanced mathematics beyond the 10th grade."[4]

In terms of the amount of time spent on science instruction after grade 6, there is also a sharp contrast between the Soviet and U.S. school programs. In the junior high schools and senior high schools in the United States, science instruction is offered through courses which are generally offered on a full-year basis. The most common science courses offered in grades 7–9 in descending order are: general science, earth science, life science, physical science, and biology. General science is the only science course offered by more than 50 percent of all the schools with grades 7–9. In grades 10–12, the most frequently offered courses are: biology, chemistry, and advanced biology.

[4]National Science Foundation and Department of Education, *Science and Engineering Education for the 1980s and Beyond*, October 1980.

Table II-3 shows U.S. public school enrollment in science courses in grades nine through twelve for 1961 and 1973. These data confirm that general science and biology are taken by more students than the other science subjects. Chemistry and physics were taken by relatively few students, and the percentage of students taking these subjects fell between the years reported. Finally, the percentage of students taking science courses in grades nine through twelve actually declined from 55.6 percent in 1961 to 44.0 percent in 1973.

Contrasting the number of hours of instruction only partially describes basic differences between U.S. and Soviet mathematics and science instruction. For instance, in a recently-prepared preliminary survey of curriculum-related materials published in the Soviet Union, the author concluded that upper secondary Soviet mathematics and science courses are comparable to the U.S. college introductory courses.[5]. The research also revealed that the Soviet Union had more and better trained mathematics and science teachers in its secondary schools than did the United States.[6]

The sharp contrast between the Soviet 10-year school curriculum and the broader American elementary-secondary education reflects in part of the general differences between the American and the European educational system, after which the Soviet Union modeled its curriculum. As opposed to the European system, however, the mathematics and science emphasis of the Soviet educational system is not tailored to selected students but to the entire school population. Thus, all students are exposed to the mathematics and science-oriented curriculum throughout their early and secondary education.

C. *Post-Secondary Educational Options*

Soviet students have four principal options upon completion of grades 1–8. (See Figure II-1 on page 18). One option is to continue general secondary education for another two years to prepare for the entrance examinations for higher educational institutions. (These two grades are considered an integral part of secondary education in the United States.) A second is to enter the specialized secondary institutions which provide the student with the final two years of secondary education plus two to three additional years in specialized post-secondary instruction.

---

[5]Nicholas DeWitt, "Current Status and Determinants of Science Education in Soviet Secondary Schools" prepared for the National Academy of Science, Washington, D.C., pp. 23–24 (April 1980).
[6]Ibid.

TABLE II-3

U.S. TOTAL ENROLLMENT AND ENROLLMENT IN SPECIFIC SCIENCE COURSES
IN GRADES 9-12 IN PUBLIC SCHOOLS: 1961 AND 1973

	Number Enrolled (In millions)		Percent of Total Enrollment	
	1961	1973	1961	1973
Total Enrollment	8.82	14.19	100.0	100.0
Total Enrollment in Science Courses	4.90	6.25	55.6	44.0
General Science	1.83	1.10	20.7	7.8
Biology	1.78	2.87	20.2	20.2
Physiology	.065*	.109	0.7	0.8
Earth Science	.076	.558	0.9	3.9
Chemistry	.745	1.03	8.4	7.3
Physics	.402	.583	4.6	4.1

Sources: Statistical Abstract of the United States, 1965, p. 119.
           Statistical Abstract of the United States, 1975, p. 126.
           National Academy of Sciences and National Research Council "The State of
           School Science." A Review of the Teaching of Mathematics, Science and
           Social Studies in American Schools, and Recommendations for Improvements,
           published as a report of the Commission on Human Resources of the
           National Research Council (1978), p. 20.

The curricula of these institutions prepare the student as a technician in one of a number of specialties which will permit him to be employed within the pertinent sector of the economy. (These are discussed in Chapter III.) A third option is vocational-technical education. The curricula for vocational-technical institutions focus on the training of skilled manpower. The system provides course work to complete the secondary educational program if the student has not already done so, as well as specific training in vocational-technical fields. Under the fourth option, students, particularly those whose academic performance has not been up to standards or who for a variety of reasons elect not to pursue the other options (for instance, youth in rural sections of the USSR), enter directly into the labor force as unskilled workers.

General secondary education has historically been the educational path for students desiring to go directly into higher education. However, the percentage of graduates of general secondary education actually admitted to higher educational establishments has been falling from 63.7 percent in 1965 to 26.2 percent in 1976.[7] Those general secondary school graduates who were not admitted to higher educational establishments entered either specialized secondary schools, sought vocational-technical training, or in some cases simply awaited another opportunity to take entrance examinations while neither in school nor in the labor force. Soviet educational specialists have been critical of students remaining outside the labor force while seeking admission to higher education, and authorities have attempted to make alternative educational options attractive. However, in the Soviet society, upward mobility within the system is dependent upon education, and a diploma from a higher educational institution opens up opportunities for mobility.

In the United States, those students who do not successfully complete their secondary education (through grade 12) generally enter the labor force. Some, however, may enter vocational training schools to acquire skills related to particular occupations. Others who fail to earn a high school diploma sometimes complete their education at adult night schools. The student who has a diploma either enters the work force or goes on to higher education to earn either a bachelor's degree or an associate degree. The educational level of the associate degree is the closest U.S. equivalent to the Soviet specialized secondary educational program.

---

[7]See Chapter IV, Table IV-4.

D. Summary

Soviet educational policy has the objective during the early and secondary years of the educational program of ensuring that the future labor force is exposed to science and mathematics in order to facilitate the Soviet goal of rapid transformation of the economy to a scientific-technical base. This goal is also consistent with the requirement for better trained and more technologically-oriented persons to fill the ranks of the Soviet armed services. However, comparative analysis of science and mathematics training in the United States and the Soviet Union should not be allowed to totally eclipse some important arguments in the United States for the well rounded, more diverse curriculum which is important to the intellectual and social development of students who will live in an interrelated, multicultural world. In addition, the mathematics/science orientation of the Soviet general curriculum does not necessarily result in a high level of competence among the total population. The evidence presented in the Soviet literature suggests that the quality of instruction is not uniformly good (of course, neither is it uniform in the United States). For instance, there is a wide disparity in the quality of instruction in urban as opposed to rural schools (it has been reported that 34 percent of schoolteachers do not have higher education, and that the majority of these teach in rural schools[8]) and in the Russian as opposed to the southern, non-Russian republics. In addition, Soviet student/teacher ratios for grades 1–10 are high by U.S. standards, with 30–40 students for each teacher, and Soviet schools still have problems with facilities, particularly laboratories. In general, however, Soviet students who finish the complete general secondary curricula receive a better preparation in science and mathematics than do their U.S. counterparts.

[8]N. Panferova, "Too Many or Too Few," *Literaturnaya gazeta* (Nov. 23, 1977), p. 13, cited in the *Current Digest of the Soviet Press*, Vol. XXIX, No. 46 (Dec. 14, 1977).

CHAPTER III

Training of Technicians

In the past, scientists and engineers worked directly with craftsmen and other skilled workers; however, as technology became more complex, it became increasingly difficult for craftsmen, who usually had a limited knowledge of science and mathematics, to work directly with scientists and engineers. A need thus developed for "technicians," persons specially trained to work with and perform some of the tasks that otherwise would be done by scientists and engineers.

The United States and the Soviet Union have similar needs for such appropriately trained support personnel for the professionally-trained science and engineering community. The demand for such technically-trained personnel has grown apace with the increasingly technical requirements of both economies. Given similar utilization requirements, the United States and the Soviet Union are both faced with providing such personnel from their respective educational systems. While utilization requirements are similar, however, the training of technicians is pursued somewhat differently. This is primarily a reflection of structural differences in the way the two countries approach education: the United States relies on diverse, pluralistic, decentralized education and the Soviet Union directs the educational system through uniform standards and centralized administration.

With manpower needs similar to those of the United States, the Soviet Union focuses on the specific production of technician-level graduates who are armed with a functional education designed to fulfill specific roles in the labor force. The Soviet Union has developed a type of education—the specialized secondary school—that provides technical-applied training at the secondary education level and is the principal source of the technical cadres that work under the direction of graduates of higher education, particularly in the engineering fields. The American educational system is much more comprehensive, offering different options in the instructional program which, although they may share a common core curriculum, allow the student a much more diverse set of choices. In the United States, there is some training specifically structured in terms of the technician role but there are differences with the Soviet system and it is not on as large a scale. The United States relies more on the manpower trained under the general high school curriculum and undergraduate training at the university level to provide the technicians required by industry, educational institutions, the government, and other employment sectors.

A. *Technician Training in the Soviet Union*

Specialized secondary institutions are a fundamental part of the Soviet system of education, organized and administered to respond to the specific demands of the national economy for technically-qualified personnel. The individuals produced by these institutions are equipped with a specialized technical education highly oriented toward industrial application at a specific position and level. The Soviet regime has emphasized this type of training in order to infuse the economy and its associated research and development institutes with a trained cadre of scientific and technical workers. Through the system of administration, direction, and support of specialized secondary schools, Soviet planning commissions can ensure that there will be a supply of technicians to fulfill the changing needs of the economy as expediently as possible. Vocational-technical schools attract a significant number of students who are trained as semiskilled workers for agriculture and industry, but these workers cannot be considered technicians in the same light as those educated at specialized secondary schools.

Specialized secondary schools provide technical training to both eighth and tenth grade graduates of general secondary schools, and offer both full-time and part-time programs. Specialized secondary schools graduate persons in a wide range of specialties. The curriculum for the eighth grade entrants involves about 75 percent technical-applied training and the program is about four years in duration. Tenth grade graduate entrants are trained in technical skills and their programs are between one-and-one-half and two-and-one-half years in duration, depending on the technical specialties.

Currently, specialized secondary schools train technicians in over 450 different specialties in 21 different specialty groups. Because of advances in science and technology and changes in production priorities, both the number of specialties taught and the content of the training at specialized secondary schools have changed over time. About 20 new specialties were added between 1965 and 1972 and in 1971–72 a number of specialties were consolidated. Official policy at the present time is directed toward providing a broader training in each specialty.

Changes in the needs of the national economy have led to shifts in enrollment by specialty over time. Table III-1 shows the number of students enrolled and graduated from Soviet specialized secondary schools by specialty from 1960 to 1978. Specialized secondary programs are likely to continue to expand, although demographic factors will probably slow the rate of increase and the number and kinds of specialties will probably change. In recent years, the specialty with the highest growth rate in enrollment and graduation has been economics, while there has been little change in enrollment and graduation in the engineering specialties. The increase in enrollment and graduation in the field of economics reflects the increased emphasis in industry on effective planning and efficient production techniques.

Specialized secondary schools are operated, financed, and managed by the Ministry of Higher and Specialized Secondary Education. The Ministry is responsible for the overall supervision of the educational process, including curricula, syllabi, textbooks, teaching methodology, and instructional methods. To facilitate the planning of specialist training, the Soviet ministries, the Council of Ministers, and the ministries of the union republics formulate lists of posts to be filled by technicians and submit annual figures to the Ministry of Higher and Specialized Secondary Education for current and long-term requirements for technicians. These lists are assessed by the Ministry of Higher and Specialized Secondary Education, which on the basis of its assessment, periodically revises and changes the emphasis of course offerings through approved academic plans and programs.

Academic plans specify the disciplines which are to be studied within a particular specialty and indicate the number of hours, courses, and semesters which are to be devoted to each particular discipline. Standards are set for examinations, evaluations, and special projects which must be fulfilled in order to obtain a qualification in the specialty.

Subjects included in the academic plan are broken into three cycles: general educational, general technical, and special applied. The general educational cycle is designed to give the student the basics of a complete secondary education, just as in the general secondary school. The cycle includes courses such as mathematics, literature, history, physics, chemistry, foreign language, and social studies—courses that would be found in the upper grades of the general secondary school program.

Courses included in the general technical cycle are determined by the general requirements of each specialty. The general technical group

TABLE III-1

U.S.S.R. ENROLLMENT AND GRADUATION FROM SPECIALIZED SECONDARY EDUCATIONAL INSTITUTIONS, BY MAJOR FIELD OF STUDY: 1960-79
(in thousands)

	1960 Enroll-ment	1960 Gradu-ation	1965 Enroll-ment	1965 Gradu-ation	1970 Enroll-ment	1970 Gradu-ation	1975 Enroll-ment	1975 Gradu-ation	1976 Enroll-ment	1976 Gradu-ation	1977 Enroll-ment	1977 Gradu-ation	1978 Enroll-ment	1978 Gradu-ation	1979 Enroll-ment	1979 Gradu-ation
Geology & Exploration of Mineral Deposits	11.8	2.5	19.9	2.5	24.8	4.8	24.3	5.7	23.9	5.6	23.1	5.9	22.5	5.9	22.1	5.6
Exploitation of Mineral Deposits	42.6	14.1	42.7	6.0	68.2	11.7	54.5	13.1	52.4	11.3	50.8	12.1	49.6	11.6	49.5	10.8
Power Engineering	98.4	15.4	180.4	26.6	218.5	44.8	194.6	47.7	199.1	37.6	198.3	43.9	196.1	44.7	193.7	45.1
Metallurgy	27.3	5.6	41.5	6.8	50.6	10.2	51.6	10.9	54.1	9.5	54.6	11.1	54.9	12.2	54.6	12.5
Mechanical Engineering and Instrument Construction	348.2	74.9	529.4	86.3	572.9	124.0	542.5	125.6	568.6	99.7	577.0	121.7	576.9	128.5	570.7	130.1
Electronics, Electrical Instrument Construction & Automation	45.5	6.8	140.2	18.7	131.5	31.8	142.6	33.8	147.9	27.8	149.9	34.3	151.3	35.2	150.4	36.3
Radio Engineering & Communications	71.1	12.5	140.0	21.8	138.7	30.9	141.7	31.6	144.6	28.7	147.6	33.0	148.2	34.6	147.0	35.8
Chemical Technology	43.5	7.0	90.9	13.4	86.2	23.3	72.6	19.0	76.7	14.6	75.7	18.4	75.4	18.4	75.2	18.6
Timber Engineering and the Technology of Woodpulp, Cellulose and Paper	28.7	6.7	39.7	6.4	46.9	8.5	48.2	10.6	48.9	10.1	48.6	11.0	47.6	11.7	46.5	11.7
Technology of Food Products	66.6	12.0	118.9	18.8	150.7	28.8	164.1	39.5	167.0	39.8	168.8	40.7	169.6	42.4	170.0	43.4
Technology of Consumer Goods Industry	59.7	9.0	102.0	17.2	109.9	24.7	108.4	24.8	111.1	22.3	113.0	24.4	113.5	25.5	113.5	26.6
Building Construction	152.0	34.2	247.7	36.2	362.7	61.9	435.3	99.8	441.0	93.7	436.6	102.6	429.0	104.8	414.1	105.6
Geodesy and Cartography	6.4	1.4	7.3	1.1	9.0	1.5	14.1	2.8	14.7	3.0	15.0	3.3	14.9	3.6	14.5	3.8
Hydrology and Meteorology	6.3	1.4	7.6	1.5	7.2	1.7	7.2	1.6	7.2	1.6	7.0	1.8	6.6	1.7	6.4	1.7
Agriculture and Forestry	292.4	67.2	497.6	68.7	601.1	120.1	645.6	142.3	661.4	144.4	670.6	150.1	678.6	157.3	675.9	161.6
Transportation	112.3	21.3	233.9	33.6	273.0	55.6	294.9	60.4	302.4	62.8	307.6	64.6	**306.5**	71.1	303.4	71.5
Economics	261.5	71.6	476.8	104.1	622.8	188.6	628.6	208.3	636.1	217.6	639.5	223.6	641.5	230.6	636.7	238.1
Public Health and Physical Culture	176.3	64.4	345.1	76.0	446.2	138.5	430.1	142.0	433.1	142.0	437.3	143.7	441.5	148.9	447.9	150.0
Art	54.6	7.5	97.6	16.3	123.6	21.2	124.6	26.9	125.0	25.9	125.8	26.2	126.1	26.8	126.3	26.9
Education	154.3	47.9	299.0	59.5	340.1	100.0	395.0	109.1	402.9	109.8	410.3	112.1	415.0	111.2	421.5	115.5
Totals	2,059.5	483.5	3,659.3	621.5	4,388.0	1,033.3	4,524.8	1,157.0	4,622.8	1,109.1	4,662.2	1,186.0	4,671.2	1,228.4	4,646.5	1,253.3

Source: The National Economy of the U.S.S.R. in 1979, Moscow "Statistika," pp. 494, 501
The National Economy of the U.S.S.R. in 1970, Moscow "Statistika," pp. 639, 647

consists of technical and specialized courses sometimes divided into two subgroups separating general technical or nonspecialized courses from courses of narrow specialization. In the specialities of industry, construction, transportation, and communications, the general technical cycle includes subjects such as drafting, technical mechanics, and electronics. The specific courses and number of hours devoted to these subjects, however, vary according to the specialty. The special applied cycle consists of courses of specific applied practical training, which determine the profile of the particular specialist. During each academic year, the student must complete two or three projects related to the particular courses of the special applied cycle, generally on the machinery and technology of production, as well as a series of courses which serve as supplementary work on machine detail.

Regardless of the specialty, academic plans for full-time specialized secondary students include no more than 36 hours per week of required theoretical studies. Great emphasis is placed on the practical experience of future technicians. Students generally spend from three to five months of their program at industrial practice assignments. After completing on-site training, the student returns to his theoretical training in the special cycle and takes examinations. After completion of the examinations, the student works on a "diploma project," which is generally required for completion of the program of studies. This work is undertaken while working at an industrial enterprise of the type in which the student will be placed upon graduation. Having finished this pre-graduation work, the student once more returns from the enterprise to the educational institution to complete his project, which he must then defend before the State Qualifying Commission. In a few specialties, such as those in some transportation, economics, and trade specialties, students are not required to complete a diploma project, but are instead required to pass examinations before the State Qualifying Commission.

While full-time specialized secondary education is considered the most effective means of training technicians, correspondence courses of study are available in most of the individual specialties. About two-thirds of all students in specialized secondary education are enrolled in full-time programs while approximately one-fourth are in correspondence programs and one-tenth are in night school. (See Table III-2.) Since the mid-1960s, the number of students enrolled in full-time programs has been increasing in proportion to the total number of specialized secondary students, with the percentage of full-time students growing from 50.2 percent in 1965 to 62.4 percent in 1978. Correspondence and night school training, however, is still valued as

TABLE III-2

U.S.S.R. FULL-TIME AND PART-TIME ENROLLMENT
IN SPECIALIZED SECONDARY SCHOOLS:
1965-1979

(In Thousands)

	1965	1970	1975	1976	1977	1978	1979
Total Students	3,659	4,388	4,525	4,623	4,662	4,671	4,646
Full Time	1,835	2,558	2,817	2,867	2,905	2,916	2,910
Correspondence	1,196	1,185	1,192	1,202	1,210	1,220	1,215
Night	628	645	516	554	547	535	521

Source: National Economy of the USSR in 1979, p. 492.

a means of providing a technical secondary education to youth who would not otherwise have the opportunity.

The organization of night and correspondence training for specialized secondary students differs considerably from regular-day training. Night school programs involve from 12 to 16 hours a week of study and cover the same disciplines as the regular day school program, but require from 6 to 12 months longer to complete the program. Night school does not cover as many areas of industrial practice and does not concentrate as heavily on the general technical and basic studies as does the full-time program. The number of disciplines covered by night programs in the general technical and general educational cycles, however, is relatively the same as that of full-time day programs, but some of the more specific, detailed subject matter is omitted, as this knowledge is generally acquired in the normal course of work for the working student.

More than 70 percent of full-time day students receive some form of financial assistance, with the level of assistance increasing proportionately with the student's level of academic excellence. Working students engaged in the night and correspondence programs receive assistance in the form of time off from work with full pay.

B. Technician Training in the United States

In the United States, technicians are defined as such by the specific employer, with the criteria ranging from the skills of a craftsman to those of a professional. Employers also vary a great deal in the levels of training they require for their technicians: some require a bachelor's degree, others require an associate's or vocational degree, and some require only a high school diploma. For this reason, there is no specific definition of a technician in general usage in the United States. The National Science Foundation has adopted the following definition of technicians employed in science and engineering for use in its statistical reports:

Technicians include all persons employed in positions which involve technical work at a level requiring knowledge in any of the fields of engineering, mathematics, physical sciences, environmental sciences, life sciences, psychology, or social sciences comparable to that acquired through formal post-high school training (less than a bachelor's degree), such as that obtained at technical institutes and junior colleges or through equivalent on-the-job training or experience. All personnel performing the duties described above should be reported as technicians even if they hold a bachelor's or higher degree.

The level of training of persons classified as technicians may range from formal four-year courses of post-secondary education to informal on-the-job training, where the individual has acquired skills while working in a technician's position. In addition, technicians are often prepared for their work through more than one type of training. Some workers are employed as technicians after having started training for another occupation, a notable example being the dropout from a professional science or engineering school. Other individuals are employed full-time as technicians while working toward professional status as a scientist or engineer. Finally, as the work of the technician becomes more complex, additional training is often required, and this training varies as much as the preemployment training.

The wide diversity in the educational background of the U.S. technician population is evident from the following data:[1] In 1972, about one-third had no post-secondary education while about one-sixth had at least a bachelor's degree, and about one-half fell somewhere between these two extremes. Of the total, 9.3 percent reported having an associate degree. Those who attended college in most cases reported having scientific or technical majors. However, in most fields, college study was not concentrated in a single major; a variety of related major fields was reported, indicating a more generalized background than is found in the scientist and engineer population.

This variety of training undergone by individuals employed as technicians in the United States makes analysis of such training more complex than in the Soviet case. The Bureau of Labor Statistics (BLS) of the U.S. Department of Labor has developed one useful classification of technician training which will be used here.[2] This classification scheme includes three broad categories, each having subcategories:

● Preemployment training

—Secondary schools

[1]Michael G. Finn, "1972 Professional, Technical and Scientific Manpower Survey," as reported in *Science and Engineering Technicians in the United States: Characteristics of a Redefined Population, 1972*, Manpower Research Programs, Oak Ridge Associated Universities, ARAU-138 (February 1978).

[2]Bureau of Labor Statisics, U.S. Department of Labor, "Technician Manpower: Requirements, Resources, and Training Needs," *Bureau of Labor Statistics Bulletin No. 1512*, pp. 33–42 (1966).

—Post-secondary schools[3]
—Industry
—Government

● Technician-related training

—Bachelor's degree programs in colleges and universities
—Armed forces

● Work experience

—Upgrading

1. Preemployment Training

Several types of secondary schools in the United States provide preemployment training for technicians. These include conventional schools (comprehensive high schools), vocational-technical schools, and technical high schools (including vocational or trade schools). Most of the training at this level is preparatory to training at a higher, post-secondary level, although a few secondary graduates are able to enter directly into a technician job.

Post-secondary institutions include technical institutes, junior or community colleges, vocational-technical schools, and extension courses offered by regular four-year colleges. While the curriculum offerings of all these schools are diverse, the typical training program is two years in length, and includes courses in science and related general topics as well as courses with specific technician job relevance (drafting, instrumentation, and so on). The emphasis in most of these schools is applied rather than theoretical, in contrast to the early courses in a professional engineering or science program.

According to Bureau of Labor Statistics estimates for 1963, the most recent available data for this particular group, about 100,000 students were enrolled in some 450 post-secondary institutions, about 100 of which were accredited by the Engineers Council for Professional Development. Three-quarters of the students came directly from secondary schools, while the other quarter entered from employment in industry or from the military. About 25,000 students graduated from preemployment training, and of these, about 16,000 took technician

[3]It should be pointed out that post-secondary schools also provide supplementary training to technicians who leave their jobs to take additional training and then return to their jobs. For example, the BLS study reports that the Engineers Joint Council found that, for graduates of two-year technician programs, 9 percent were returning to a job after graduation.

jobs. A large number of the remainder (unreported) continued their schooling in a four-year college.[4]

More recent data specifically dealing with technicians have not yet been compiled. On the basis of data collected by the National Center for Education Statistics, however, BLS reports recent data for post-secondary vocational education, which is described as "intended for persons who have completed or left high school and includes those who are enrolled in programs leading to an associate or other degree below the baccalaureate."[5] The number of individuals who completed associate degrees in science and engineering in 1977–78 is shown in Table III-3, and the number of students enrolled and completing both public and private post-secondary vocational training in technical specialties is shown in Table III-4. It should be noted that these numbers do not include individuals who attained technical education through night school programs, employment training, or military training programs.

According to the 1963 Bureau of Labor Statistics study, industry training for technicians is used primarily when workers with suitable educational and other qualifications cannot be found. Industry typically recruits candidate trainees from among the more promising workers already on the payroll. Programs offered are generally combinations of formal academic course work and on-the-job activities. The academic work may take place within the firm or in outside educational institutions of various kinds. Some training programs are formal registered apprenticeships. Most such programs are for blue-collar trades rather than technicians, but the lines are becoming less distinct than in the past. Most of the technician apprenticeships are for draftsmen. Bureau of Labor Statistics estimates for 1963 found that over 20,000 individuals were trained in these types of industry programs. In addition, a large number of technicians (unreported by BLS) were given industry training to acquire needed new skills.

Government preemployment programs are funded under a variety of statutes. Title VIII of the National Defense Education Act of 1958 provided training for highly skilled technicians. The Vocational Education Act of 1963, which remains in effect, made this type of training permanent. The Manpower Development and Training Act of 1962 is aimed at trainees in technical and other occupations in localities where there is a demonstrated labor shortage.

[4]Bureau of Labor Statistics, op. cit.
[5]"Occupational Projections and Training Data," 1980 Edition, U.S. Department of Labor, *Bureau of Labor Statistics Bulletin No. 2052,* p. 10 (September 1980).

TABLE III-3

NUMBER OF U.S. INDIVIDUALS COMPLETING ASSOCIATE DEGREES
IN SCIENCE AND ENGINEERING: 1971-72 to 1977-78

	1971-72	1972-73	1973-74	1974-75	1975-76	1976-77	1977-78
Mechanical and Engineering Technologies	44,145	34,781	37,631	40,775	45,169	49,249	51,200
Natural Science Technologies	9,418	9,242	11,496	12,966	13,316	15,534	15,980

Source: Occupational Projections and Training Data, 1980 Edition, U.S. Department of Labor, Bureau of Labor Statistics, Bulletin 2052, p. 107 (September 1980).

TABLE III-4

NUMBER OF U.S. INDIVIDUALS ENROLLED AND COMPLETING
PUBLIC AND PRIVATE POSTSECONDARY VOCATIONAL TRAINING
IN TECHNICAL SPECIALITIES: 1977-78

	Enrollments	Completions
Public and Private Vocational Education, Technical Occupations	559,826	114,554

Source: Occupational Projections and Training Data, 1980 Edition,
         U.S. Department of Labor, Bureau of Labor Statistics
         Bulletin 2052, pp. 111-113 (September 1980).

2. Technician-Related Training

There are many similarities between the four-year college science or engineering program leading to a bachelor's degree and the post-secondary training given to technicians, as the duties of both professionals and technicians are similar in nature although not in depth. However, the emphasis on theory is much greater in the four-year program. Consequently, engineering students educated in the first two years of a four-year sequence are unlikely to have the skills required to perform the work of a technician. Industry, however, employs as technicians many individuals with a bachelor's degree in science or engineering, but primarily for training preparatory to reclassification as a professional.

A number of colleges and universities now offer four-year programs that grant bachelor's degrees in technology. These programs are less theoretical and less mathematical than their professional counterparts and are more hardware and process-oriented. At the same time, graduates of these programs typically receive a broader education than technicians, with more emphasis on the humanities and technical disciplines and greater depth in studies of mathematics and science. Graduates, termed "technologists," are trained to be placed in teams working with both technicians and professional engineers and scientists.

The armed forces play an important role in technician training. Typically, the instruction is specific to a given set of tasks and equipment, and includes classroom, laboratory, and on-the-job components. While there is inadequate statistical information on the number of technicians who use their military training to obtain civilian jobs, the number is believed to be quite small according to the Bureau of Labor Statistics. When they do switch to industry, additional training is usually required. However, there is some evidence from a 1972 Postcensal Survey prepared by the National Science Foundation that suggests that military training may be more pertinent to the work of technicians than the BLS study would imply.[6] In the Postcensal Survey it was found that over 15 percent of "engineers," prior to screening for redefinition purposes, reported having military training related to their civilian work.

The Postcensal Survey classified 329,000 individuals as falling into the technician category.[7] A subsequent analysis of the data pertaining to this set of individuals found that 61,500 persons (18.7 percent of

---

[6]*The 1972 Scientist and Engineer Population Redefined,* 2 Vols., National Science Foundation, NSF 75-313 and NSF 75-327 (1975).
[7]Including computer programmers. See Chapter VI, Section B, for a discussion of the methodology used in this study.

the total) reported military training applicable to civilian occupations.[8] Of these persons, 73.4 percent reported that their military training experiences were useful in their occupations.[9]

3. Work Experience

According to the study of the characteristics of the redefined technician population from the 1972 Postcensal Survey,[10] approximately 68 percent of the technicians reported participating in some form of supplemental training outside of their formal educational programs. Participation in such training ranged from 50 percent for biological and agricultural technicians and physical science technicians to more than 80 percent for computer programmers and electrical and electronic technicians.

In the aggregate, the incidence of training does not vary systematically with the level of education, as is shown in Table III-5. However, these aggregate figures undoubtedly mask a variety of different forms and uses of training related to the level of formal education. Different forms of training can generally be classified as occurring before or after employment. Military training almost always precedes employment for which the training is used. On-the-job training, courses at an employer's training school, and apprenticeships usually follow employment. Correspondence, extension, and adult education courses—some related to employment, some not—are taken for a variety of reasons and can either precede or follow employment.

In conclusion it should be noted that problems in quantifying persons receiving or completing training as technicians in the United States were noted in the recommendations made to the White House by the National Science Foundation and the Department of Education in their report, *Science and Engineering Education for the 1980s and Beyond,* released in October of 1980. This report concluded that there is a primary need for an improved information base on technician training in the United States and recommends that "the National Science Foundation conduct a comprehensive survey of industry to obtain data about technician needs, training and utilization and that information regularly collected by the Bureau of the Census and Bureau of Labor Statistics be examined to determine its quantity and quality."[11]

---

[8]Finn, op. cit., p. 30.
[9]Ibid., p. 30.
[10]Ibid., p. 30.
[11]*Science and Engineering Education for the 1980s and Beyond,* National Science Foundation and Department of Education, pp. xxxiii-xxxv (October 1980).

TABLE III-5

PERCENTAGE OF U.S. TECHNICIANS REPORTING SUPPLEMENTAL
TRAINING, BY LEVEL OF EDUCATION: 1972

Level Of Education	Percent Reporting Training
Some secondary	58.7
Graduate, secondary	66.8
1-2 yr. college, no degree	68.9
3+ yr. college, no degree	67.3
Associate degree	65.1
Bachelor degree	68.1
More than bachelor degree	61.8
All levels combined	67.6

Source: Michael G. Finn, Science and Engineering Technicians
         in the United States: Characteristics of a Redefined
         Population, 1972, Manpower Research Programs,
         Oakridge Associated Universities, ORAU-138, p. 25
         (February 1978).

CHAPTER IV

Higher Education

The major contrast between the U.S. and Soviet systems of higher education reflects the general difference in educational philosophy in the two countries, with the United States placing more emphasis on general education with a significant amount of individual choice and the Soviet Union emphasizing functional training to provide workers to fill the needs of the economy. Professional specialization in the USSR is considered to be far more pronounced than in almost any other country in the world. In contrast to the liberal arts and other nonprofessionally oriented programs common in American colleges and universities, all higher educational programs in the Soviet Union are professionally oriented, involving a degree of specialization even greater than that in the functionally-oriented courses of study in professional schools in the United States.

The student's field of study at a Soviet higher educational establishment is designated by the term "specialty." Students choose a specialty at the time they apply for admission to a higher educational establishment, and, once admitted, follow a rigidly defined program of study preparing them for a professional occupation in that specialty. Currently, there are about 360 specialties offered by Soviet higher educational establishments. These are combined in 22 different specialty groups, shown in Table IV-1. A complete list of the 360 specialties for students' fields of study at Soviet higher educational establishments is presented in Appendix A. Mechanical engineering and instrument construction, for example, is one of the 22 specialty groups; within the mechanical engineering group, there are 46 different specialties (such as peat mining machinery and outfits, machinery and technology of casting processes, boiler construction, and hydro-aerodynamics) which constitute the student's actual field of study. Over 200 of the specialties offered in Soviet higher educational establishments are in engineering-industrial fields.

The extremely narrow specialization characteristic of Soviet higher

TABLE IV-1

LIST OF SPECIALTY GROUPS AND NUMBER OF SPECIALTIES IN U.S.S.R.
HIGHER EDUCATIONAL INSTITUTIONS: 1972

	Number of Specialties
Specialty Groups	
Geology and exploration of mineral deposits	7
Exploitation of mineral deposits	11
Power engineering	12
Metallurgy	9
Machine-building and instrument construction	46
Electrical machine building and electrical instrument construction	35
Radio engineering and communications	7
Chemical technology	26
Timber engineering and technology of wood processing, cellulose and paper manufacturing	4
Technology of food products industry	16
Technology of consumer goods industry	13
Construction	16
Geodesy and cartography	4
Hydrology and meterology	5
Agriculture and forestry	13
Transportation	15
Economics	39
Law	3
Public health and physical culture	6
University specialties	32
Specialties in pedagogical institutes and higher educational institutions of culture	18
Art	27
	364

Source: U.S.S.R. Ministry of Higher and Secondary Specialized Education,
List of Specialities and Specializations of USSR Higher Educational
Institutions, Moscow, 1972.

educational training has often been faulted for its failure in providing scientists with the ability to master new knowledge, assimilate new research methods, and cope with technological change. In addition, such narrowly specialized training is highly susceptible to obsolescence and has for this reason been the subject of frequent controversy among Soviet education specialists. Most recently, the need for more flexible curricula and greater emphasis on general theoretical background in the training of "broad-spectrum" specialists was called for in a decree on higher education issued in July 1979. Despite the debate that has often surrounded the issue of whether higher education should train narrow specialists or personnel with a wider range of knowledge, the basic pattern of higher education in the Soviet Union has remained fundamentally the same as it was in the 1960s.

Specialization in U.S. higher educational institutions generally commences after the completion of a core curriculum of course work designed to give the student a broad background in a multiplicity of disciplines. These courses, along with the core courses of the major field and elective courses, compose the requirements for graduation. The composition of these requirements varies with the program of study, the discipline, and the variety of ways that study materials are organized into courses in various institutions. This generalized initiation into all academic disciplines offers the student a great deal of latitude in deciding on a field of specialization. The decision can often be deferred until the third year of college, at times even later. Programs of specialization are generally much less structured than those found in the USSR, with a significant number of elective courses left to the discretion of the student.

A. Types of Higher Educational Institutions and Degrees

Soviet higher educational establishments are essentially of two types: universities and institutes. Universities generally offer programs of instruction in a variety of fields, primarily in the natural and social sciences, while institutes usually concentrate on a single area of related specialties specifically oriented toward a given sector of the economy, agriculture, or medicine. In addition, a special type of instruction referred to as the "factory higher technical education school" (Zavod VTUZ), trains technological engineers on the actual premises of large industrial enterprises.

The training provided by universities is generally somewhat broader and of a more theoretical nature than that provided by institutes. From among the 22 specialty groups of higher educational programs in the

Soviet Union, a group referred to as "university specialties" is singled out as a discrete category. Specialties in this category, which include physics, mathematics, biology, etc., as well as social sciences and humanities, appear to have a greater similarity to fields of study pursued by students in U.S. colleges and universities than do many of the more narrow, applied specialties offered in Soviet institutes. Even in the university specialties, however, higher education in the Soviet Union tends to be more functionally oriented than in the United States, although it is in this case designed with a view toward professional occupations in basic research or teaching as opposed to the more narrowly applied industrial orientation of most other specialties in the USSR.

As shown in Table IV-2, out of a total of 856 higher educational establishments in the USSR in 1975, there were 63 universities. Over the last three decades, Soviet universities have fairly consistently accounted for about 11 or 12 percent of the total number of graduates from higher educational establishments. While the percentage of graduates from higher educational establishments that have been trained in universities is relatively small, university training is generally considered to be qualitatively superior to training received in institutes.

Students are given a diploma upon successful completion of the academic program for a particular specialty at a higher educational establishment in the Soviet Union, as opposed to receiving an academic degree as in U.S. colleges and universities. Academic degrees in the Soviet Union are awarded only for completion of postgraduate work beyond the basic higher educational program.

In the United States, the various higher educational institutions can be distinguished with respect to the content and extent of their academic offerings as universities, four-year colleges, or two-year colleges (often called junior or community colleges). They can also be distinguished with respect to their control and support as either public or private.

Except for a handful of institutions, chiefly the military academies or postgraduate schools associated with the military services, which are federally controlled, U.S. college-level public institutions either are part of the higher educational systems of the states or are under the control of municipalities or local school or special districts set up especially for this purpose. More than half of the colleges and universities, however, are privately endowed and controlled.

The U.S. two-year colleges are widely distributed throughout the states. They are generally local or community institutions that provide an opportunity for students to obtain two years of college training while residing at or near home and to defer seeking admission to a four-year

TABLE IV-2

U.S.S.R. INSTITUTIONS OF HIGHER EDUCATION: 1960-1975

	1960	1965	1970	1971	1972	1973	1974	1975
All Higher Educational Institutions								
Number	739	756	805	811	825	834	842	856
Enrollment (in thousands)	2396.1	3860.6	4580.6	4597.5	4630.2	4671.3	4751.1	4854.0
Universities								
Number	40	42	51	52	58	60	63	63
Enrollment (in thousands)	249.0	401.2	503.5	509.8	539.0	543.9	561.3	565.9
University Enrollment as a percentage of total higher educational institution enrollment	10.4	10.4	11.0	11.1	11.6	11.6	11.8	11.7

Source: National Education, Science and Culture in the U.S.S.R., Moscow "Statistika," 1977, pp. 213, 218;
National Economy of the USSR in 1970, p. 637;
National Economy of the USSR in 1975, p. 677;
National Economy of the USSR in 1979, p. 492.

college until ready for the third (junior) college year. The two-year schools grant Associate of Arts (A.A.) or Associate of Science (A.S.) degrees. The completed curriculum is the equivalent of the first two years at a four-year college or university. The two-year colleges also provide some opportunities for vocational training and grant Associate of Applied Science (A.A.S.) degrees for technicians or midmanagement type jobs. (See Chapter III.) In addition, they often offer programs for adult continuing education at the collegiate level.

U.S. four-year colleges provide a full academic program leading to a baccalaureate degree, either a Bachelor of Arts (B.A.) degree or a Bachelor of Science (B.S.) degree. Although many of these colleges grant master's degrees, few offer doctorates. Universities provide a wide range of training, offering all levels of undergraduate and advanced degrees, and often include affiliated professional schools such as for law and medicine.

Table IV-3 shows the number of institutions of higher education in the United States, by type of control and size of enrollment for the fall of 1969, the fall of 1974, and the fall of 1978. In 1978, there were 3,131 institutions of higher learning in the United States, of which 160 were universities, 1,781 were four-year colleges, and 1,190 were two-year institutions. As indicated by the table, the total number of institutions of higher education increased in the ten-year period, most of the increase taking place in public two-year colleges. Enrollment in the two-year colleges by 1978 represented more than one-third of the total enrollment in all colleges and universities. There is a wide disparity between the United States and the Soviet Union in terms of the number of higher educational institutions, with the United States having more than twice as many institutions as the USSR. This is largely due to the great number of small private colleges in the United States.

B. Admissions to Higher Educational Institutions

Students who desire to enter Soviet higher educational establishments must have completed their secondary education, whether in general educational, vocational-technical, or specialized secondary schools. Students are chosen competitively, on the basis of competitive entrance examinations as well as marks in secondary school. In contrast to the standardized national entrance examinations that often serve as extremely important admissions criteria for many U.S. colleges and universities, Soviet entrance examinations are exclusively of the achievement-test rather than aptitude-test variety (the latter were officially banned in 1936). As individual institutions formulate their own examination questions, they can vary their own standards of admission.

TABLE IV-3

U.S. INSTITUTIONS OF HIGHER EDUCATION, BY TYPE OF CONTROL, AND SIZE OF ENROLLMENT:
FALL 1969, FALL 1974, AND FALL 1978

Control of Institution and Size of Enrollment	All Institutions		Universities		All Other 4-Year Institutions		2-Year Institutions	
	Number	Enrollment	Number	Enrollment	Number	Enrollment	Number	Enrollment
Fall 1969								
Public and private institutions	2,525	7,916,991	159	2,879,847	1,480	3,094,819	886	1,942,325
Public institutions	1,060	5,839,719	94	2,181,871	332	1,839,525	634	1,818,323
Private institutions	1,465	2,077,272	65	697,976	1,148	1,255,294	252	124,002
Fall 1974								
Public and private institutions	2,747	10,223,729	158	3,231,923	1,586	3,680,259	1,003	3,311,547
Public institutions	1,214	7,988,500	93	2,514,195	354	2,279,502	767	3,194,803
Private institutions	1,533	2,235,229	65	717,728	1,232	1,400,757	236	116,744
Fall 1978								
Public and private institutions	3,131	11,260,092	160	2,780,729	1,781	4,451,222	1,190	4,028,141
Public institutions	1,472	8,785,893	95	2,062,295	455	2,849,908	922	3,873,690
Private institutions	1,659	2,474,199	65	718,434	1,326	1,601,314	268	154,451

Note: In the above tabulation a branch campus is not counted as a separate institution but is considered to be a part of the parent institution. Enrollment includes students whose programs of study are creditable toward a bachelor's or higher degree and also students in 1-, 2-, or 3-year undergraduate programs not creditable toward a bachelor's degree but designed for immediate employment or to provide general education.

Source: U.S. Department of Health, Education and Welfare, National Center for Education Statistics, Digest of Education Statistics 1970, p. 85, Digest of Education Statistics 1975, p. 98, and Digest of Education Statistics 1980, p. 109.

Depending on the higher educational establishment, examinations are required in three to five subjects from among the following: Russian language and literature, mathematics, physics, chemistry, history, geography, foreign languages.

Annual quotas on admissions for each specialty within each higher educational establishment are determined by the central government on the basis of a demand schedule for various specialists in the national economy. Every application to a higher educational establishment must be accompanied by the applicant's personal documents and must specify the specialty for which he is seeking to gain admission. Thus, only one application can be made at any one time. Entrance examination results are published at the end of August, at which time students who have not been admitted may submit applications to establishments that are known to have unfilled places in specific specialties. Unsuccessful applicants generally wait until the following year, however, before reapplying for admission to higher educational programs.

Taking the USSR as a whole, quotas probably function as a more important determinant of the overall academic ability of all students admitted to higher educational establishments than do entrance examinations, because at those higher educational establishments that receive relatively few applications compared with the number of places available, the less qualified applicants are admitted so as to fill the admissions quotas. While there is also a wide variation in the academic ability of students admitted to different colleges and universities in the United States, students who wish to enter higher educational establishments generally apply to a number of institutions, thus permitting any given institution to select its most qualified applicants and the individual to choose from among those institutions to which he is admitted. Without standardized national entrance examinations to show the relative ability of different applicants, and with students permitted to apply to only one higher educational establishment at any given time, the Soviet system probably results in the denial of access to higher education to at least some of the more qualified students throughout the country as a whole, despite the attempt to ensure that the best students are admitted to higher educational establishments and eventually to graduate training.

In the United States, any person who desires to obtain a college education can apply for admission to the college or colleges he desires to attend. Several factors are usually involved in a student's choice of a college: the type of program he desires to pursue, the location, size, enrollment, student/faculty ratio and tuition charges and other expenses of the college or university, greater familiarity with certain institutions because of parents or other relatives having attended them in the past, the religious affiliation of the institution, etc. As the decision as to whether or not to admit any particular applicant is left to the discretion of the individual institutions, another factor which generally plays a significant role in a student's choice of colleges is his estimation of the likelihood of his being granted admission, given his academic credentials and the admissions criteria of the institution. Because admissions to colleges are generally competitive and a student is not assured of being granted admission to any specific college of his choice, to insure that he is admitted to at least one institution, a student often applies to several, at least one of which he feels will find him an acceptable candidate.

Within the constraints imposed by faculty and space limitations, U.S. colleges generally attempt to admit those students who will most profit from their educational facilities and be able to meet the established academic standards of the school. As a result, the criteria for admissions vary markedly from one college to another. Many colleges have selective, highly competitive admissions standards, sometimes accepting as few as one of every ten applicants. Some four-year colleges and many two-year colleges have open admissions policies, requiring only that a student have graduated from an accredited high school or attained a high school equivalency certificate.[1] Some state systems of higher education, such as the California system, include several types of institutions with different admissions requirements.

Most U.S. colleges and universities evaluate several criteria in deciding which applicants are likely to have the ability to do creditable work and eventually earn a degree from their institution. Significant among these are the high school record or transcript showing courses completed and the grade earned in each course, the applicant's rank in his graduating class, and recommendations from teachers, department chairmen, guidance counselors, and school principals. Because of wide variations in the academic standards of different secondary schools, a large number of colleges emphasize scores on standardized national entrance examinations, such as the Scholastic Aptitude Test and the American College Testing Program's assessment tests, as well as the high school record in evaluating an applicant for admission. In addition to factors demonstrating an applicant's academic ability, colleges and universities frequently take into consideration a number of nonacademic factors, such as leadership potential, athletic activities,

---

[1] By 1970, practically all community colleges had adopted some form of open admissions policy.

or impressions made at personal interviews, which they believe are relevant to an applicant's success at their institution.

Data on U.S. and Soviet secondary school graduates entering and completing schools of higher education for 1960 to 1978 are shown in Table IV-4. While all students who have completed a secondary education are eligible to apply for admission to higher educational establishments in the Soviet Union, the general secondary schools are essentially the training ground for higher education. Accordingly, data in Table IV-4 are based on the percentage of general secondary school graduates admitted to higher educational establishments. As the percentage is based on the number of graduates of general secondary schools only, while the admissions figures include graduates of specialized secondary and vocational-technical schools as well as some students who graduated from secondary schools in preceding years, the percentage of admissions is probably somewhat exaggerated. The percentage of secondary school graduates admitted to higher education has been consistently declining since 1965, so that in 1979 only about one out of every four secondary school graduates was admitted to a higher educational establishment. In the United States, by contrast, from 60 to 75 percent of all high school graduates entered higher educational institutions in the 1970s. On the other hand, while the United States has a greater proportion of secondary education graduates entering higher educational institutions than does the Soviet Union, attrition rates are much higher. About 80 percent of those admitted to higher educational establishments in the Soviet Union complete their undergraduate education and receive a diploma, whereas only about 55 percent of those students who enroll in U.S. colleges go on to receive their bachelor's degrees.

C. Full-Time and Part-Time Programs

There are three basic types of instruction programs offered in Soviet higher educational establishments: regular day, evening, and extension-correspondence. The latter two involve part-time study and are frequently comprehended under the term "without time-off from production." Evening and correspondence programs, while employing identical teaching methods as full-time programs, differ in intensity and amount of time devoted to instruction. The quality of part-time programs, especially of the extension-correspondence type, has been questioned because of high dropout rates, excessive absenteeism, and probably inferior instruction. One U.S. analyst states categorically that the engineers trained in such programs "would not be considered to

TABLE IV-4

U.S. AND U.S.S.R. SECONDARY SCHOOL GRADUATIONS AND ENTRANCE AND COMPLETION OF HIGHER EDUCATION: 1960-1979

U.S.

Year of High School Graduation	High School Graduates (Thousands)	First Time College Students (Thousands)	First Time College Students as a Share of High School Graduates (Percent)	Year of Graduation from Higher Education	Graduates from Higher Education (Thousands)	Graduates from Higher Education as a Share of Entrants (Percent)
1960	1,864	923	49.5	1964	502	54.4
1965	2,665	1,442	54.1	1969	770	53.4
1970	2,896	1,780	61.5	1974	1009	55.9
1971	2,944	1,766	60.0	1975	988	55.9
1972	3,008	1,740	57.8	1976	998	57.4
1973	3,043	1,757	57.7	1977	993	56.5
1974	3,081	1,854	60.2	1978	998	53.8
1975	3,140	1,910	60.8	1979	NA	NA
1976	3,154	2,377	75.4	1980	NA	NA
1977	3,154	2,432	77.1	1981	NA	NA
1978	3,147	2,422	77.0	1982	NA	NA
1979	NA	NA	NA	1983	NA	NA

U.S.S.R.

Year of Graduation from General Secondary School	General Secondary Graduates (Thousands)	Higher Education Admissions (Thousands)	Higher Education Admissions as a Share of Secondary School Graduates (Percent)	Year of Graduation from Higher Education	Graduates from Higher Education (Thousands)	Graduates from Higher Education as a Share of Entrants (Percent)
1960	1,055	593	56.2	1965	404	68.1
1965	1,340	854	63.7	1970	631	73.9
1970	2,581	912	35.3	1975	713	78.2
1971	2,708	920	34.0	1976	735	79.9
1972	2,886	930	32.2	1977	752	80.9
1973	3,087	938	30.4	1978	772	82.3
1974	3,374	963	28.5	1979	790	82.0
1975	3,564	994	27.9	1980	NA	NA
1976	3,873	1,013	26.2	1981	NA	NA
1977	4,101	1,017	24.8	1982	NA	NA
1978	4,162	1,026	24.7	1983	NA	NA
1979	4,030	1,043	25.9	1984	NA	NA

Sources: Digest of Educational Statistics 1980, pp. 165, 166, 175;
Statistical Abstract of the United States 1973, pp. 130, 137;
Statistical Abstract of the United States 1979, pp. 159, 168, 160;
National Economy of the U.S.S.R. in 1970, p. 644;
National Economy of the U.S.S.R. in 1975, pp. 670, 683;
National Economy of the U.S.S.R. in 1979, pp. 488, 499.

be professionally trained engineers in the United States."[2] The advantages of this type of education to Soviet planners is that it allows the channeling of a large number of young people toward the pursuit of higher education while not removing them from the labor force. It also allows for a substantial reduction in the cost of education to the state, as students engaged in part-time programs are for the most part self-supporting.

At the end of the 1950s and early 1960s, evening and correspondence education had become the predominant means of training specialists in Soviet higher educational establishments, accounting for 51.8 percent of total enrollment in the 1960-61 academic year and 59.0 percent by 1965-66. (See Table IV-5.) In the mid-1960s, however, full-time instruction was given increased emphasis, so that by 1979, enrollment in evening and correspondence programs had fallen to 43.5 percent of total enrollment, and the ratio of first-year part-time students had decreased to about 40 percent of the total. Data for 1975 indicate that in terms of major fields of study, part-time enrollment was higher in the humanities and social sciences (42.8 percent) than in engineering (36.7 percent), physical and life sciences and mathematics (35.0 percent), or agriculture (33.1 percent). (See Table IV-6.) Soviet sources admit that this increased emphasis on full-time instruction beginning in the mid-1960s was at least in part in recognition of qualitative advantages of full-time educational programs.

The United States offers very few higher educational programs that are at all comparable to the Soviet extension-correspondence type programs.[3] Although many students obtain their higher education through part-time study, the distinction between part-time and full-time students in the United States is made only on the basis of the number of courses in which a student is enrolled during any particular semester. The overall requirements for obtaining a degree are the same for part-time as for full-time students, and part-time students are not singled out to attend different classes with different instructors from full-time students. Thus, the level of qualification of individuals who obtain a degree through part-time study does not differ from that of individuals who obtain a degree through full-time study.

[2]David W. Bronson, "Scientific and Engineering Manpower in the USSR and Employment in R&D," in *Soviet Economic Prospects for the Seventies,* Joint Economic Committee, U.S. Congress, p. 564 (June 1973).
[3]Exceptions include extension and correspondence courses offered by federal and state departments and agencies.

TABLE IV-5

USSR FULL-TIME AND PART-TIME ENROLLMENT AND GRADUATION
FROM HIGHER EDUCATIONAL INSTITUTIONS: 1960-1979

	1960	1965	1970	1975	1976	1977	1978	1979
Total (in thousands)								
Enrollment	2,396	3,861	4,581	4,854	4,950	5,037	5,110	5,186
Graduation	343	404	631	713	735	752	772	790
Regular Day (in thousands)								
Enrollment	1,156	1,584	2,241	2,628	2,711	2,789	2,861	2,932
Graduation	229	225	335	433	448	462	479	493
Evening (in thousands)								
Enrollment	245	569	658	644	650	652	653	653
Graduation	15	44	82	80	82	85	85	85
Correspondence (in thousands)								
Enrollment	995	1,708	1,682	1,582	1,589	1,596	1,596	1,601
Graduation	99	136	214	200	204	205	208	212
Percent Part Time								
Enrollment	51.8	59.0	51.1	45.9	45.2	44.6	44.0	43.5
Graduation	33.2	44.6	46.9	39.3	38.9	38.6	38.0	37.6

Source: National Economy of the U.S.S.R. in 1970, pp. 637, 645;
National Economy of the U.S.S.R. in 1975, pp. 677, 684;
National Economy of the U.S.S.R. in 1979, pp. 492, 499.

TABLE IV-6

U.S.S.R. FULL-TIME AND PART-TIME GRADUATIONS FROM HIGHER EDUCATIONAL
INSTITUTIONS, BY MAJOR FIELD OF STUDY: 1975

	Thousands				Percent Part Time
	Total	Day	Evening	Correspondence	
Engineering	272.0	172.2	53.6	46.2	36.7
Agriculture	53.9	36.0	–	17.8	33.1
Physical and Life Sciences and Mathematics[1]	(44.9)	(29.2)	(4.8)	(10.9)	(35.0)
Other	342.6	195.8	21.4	125.4	42.8
Total	713.4	433.3	79.7	200.4	39.3

[1] Figures are estimates based on two-thirds of graduations in "university
specialties" to which graduations in geology-prospecting, geodesy-
cartography, and hydrology-meteorology have been added to approximate
U.S. definitions.

Source: Calculated on the basis of data contained in USSR: Trends and
Prospects in Educational Attainment, 1959-85, National Foreign
Assessment Center, ER 79-10344, p. 30 (June 1979).

D. Course of Study

In the Soviet Union, an academic program, specifying the sequence of courses, the distribution of theoretical and practical work, the length of time of instruction, examinations, and so forth, is formulated for each specialty by the central government, although individual higher educational establishments have authority to introduce minor changes. Academic programs are designed to equip the student to perform a given occupational job depending on the specialty. The length of training from the time of entering the higher educational establishment to receipt of the diploma is from four to five-and-one-half years. Most of the science and engineering specialties involve four-and-one-half to five years of training.

Academic programs generally contain from 40 to 50 different courses or subjects which the student must complete. These subjects are organized into cycles: socioeconomic, general scientific, and specialized. In the higher educational institutes in the engineering-industrial branch, a general engineering cycle is included in addition to the other three cycles.

General theoretical disciplines, including mathematics, physics, and chemistry, are covered during the first three to three-and-one-half years of the higher educational program. Following the first phase of higher education, more intensive study directed toward the narrow field of specialization is undertaken. In the academic program for the physics specialty, for example, the general scientific cycle includes courses in mathematics, general physics, astronomy, theoretical physics, methods of solving applied problems on a computer, and the basics of radio electronics. The specialized cycle includes courses such as low temperature physics, molecular physics, optics, and spectroscopy. The general scientific cycle in electrical engineering includes courses such as industrial safety and the organization, planning, and control of industrial enterprises, as well as courses in chemistry, mathematics, physics, theoretical mechanics, and theoretical foundations of electronics. The specialized cycle includes courses such as electrical materials, electrical measurement, industrial electronics, computers, high voltage engineering, and design of power stations. Students are also required to take courses in the humanities and social sciences, such as political economy, theory of scientific communism, philosophy and history of the Communist Party of the Soviet Union, and to study foreign languages (English, French, German, or Oriental, depending on the specialty). During the latter years of the academic program, students are allowed some choice in the selection of courses.

During the last half-year of the higher educational program, students prepare for state examinations or for defense of a diploma project (similar to a thesis) before the state examination board. In the technical higher educational institutions, diploma projects are more common than state examinations.

The instruction time in most Soviet higher educational programs is usually one-and-one-half to two times as great as that in U.S. colleges and universities. Lectures account for about 50 percent of the training time. However, in the junior and senior year, a great deal of emphasis is placed on practical knowledge obtained through independent work on seminar projects or courses in which the student is required to work on a project theme that often involves laboratory research.

Instruction in factory higher technical education schools (Zavod VTUZy) that have been established at various large industrial enterprises involves a strong emphasis on practical, applied training in technological engineering specialties related to the production process of the enterprise at which the Zavod VTUZ is organized. The period of study ranges from five-and-one-half to six years, and involves a combination of theoretical study and actual production work throughout the entire course of study. Instruction takes place in auditoriums, offices, and laboratories of the Zavod VTUZ, as well as in shops, laboratories, and studios of the enterprise itself.

After graduation from a higher educational institution, students are assigned to jobs at enterprises or institutes where they are required to work for three years. Some students with particularly good undergraduate records are allowed to take entrance examinations for graduate study after two years of work. During the first year at the enterprise or institute, new graduates of higher educational institutions undergo a traineeship, or probationary period, in order to gain practical and organizational skills in their specialties.

In the United States, analysis of undergraduate curricula and graduation requirements in institutions of higher education is complicated by the great variability in the way study materials are organized into courses and in the systems used to assign values or weights to individual courses in relation to the amount of effort required for completion. The U.S. system measures progress toward graduation in terms of courses satisfactorily completed. Graduation requirements depend on the student's major field of study and include required courses, electives to be chosen from various categories relevant to the major field, others to ensure breadth of study, and finally general or unrestricted electives to complete a prescribed number of credits. These requirements vary from one college to another.

It is usually after completion of the first two years of general studies in the arts and sciences that students declare the major field of study that they wish to pursue in their next two years. Introductory courses taken in the first two years often influence students' decisions, and tentative decisions made earlier may have changed by this time. If the student is still undecided, he may confer with school counselors and/ or take aptitude tests that may help him determine his field of choice. Counselors can also provide information on employment opportunities, salary ranges, and so on. From this point on, because of the specialized nature of the last two years of study, it becomes more difficult for a student to change his field of study, although it is not uncommon. Behind all the diversity, several constraints tend to be binding: (1) the requirements set by various professional accreditation and certification associations, (2) the need to prepare students adequately for transferring to another institution for completion of the last two years of an undergraduate program or for commencing graduate study and, more generally, (3) the need to achieve a measure of success in placing terminating students in gainful employment.

Because the two-year colleges are often used as stages in a four-year educational program, their offerings must permit completion of the requirements that four-year institutions impose on their students before admitting them to the upper division (the last two of the four-year curriculum)—that is, to third-year standing. This means that standard introductory courses required for a major field must be successfully completed in addition to general requirements in mathematics, natural sciences (other than in the major field of study), social sciences, and humanities. The two-year degrees generally require courses in a major field roughly equal to one-third of the total of courses required in that major in order to obtain a bachelor's degree.

The four-year colleges tend to permit a slightly greater degree of specialization and diversity in their lower division courses (those intended to be taken in the first two years) than is often possible in the two-year colleges. The two-year colleges tend to make greater use of general introductory courses. For example, one introductory college physics course might be designed to accommodate a variety of students including those planning a physics major, others preparing for secondary school teaching and selecting physics as a teaching minor (an area of competence for principal teaching assignments), and still other students desiring the course merely to satisfy a generalized science or breadth requirement. In the four-year schools, a different course may be offered to meet the needs of each group of students. The general survey course in physics for students who are not science majors will avoid mathematics and will be adjusted to be comprehensible to the intelligent layman. The student majoring in physics, on the other hand, will be required to take a more specialized and technical course and be responsible for first acquiring the necessary mathematics or other prerequisites for the physics course. Engineers may be required to take a specially designed course in physics stressing engineering mechanics and other engineering applications.

In general, the students in lower division work in any department all tend to follow the same sequence of courses, but follow individual progressions in the last two years. These differences are advantageous because they permit students to tailor their full program to their educational goals, e.g., to selected graduate specialties or to intended postgraduate vocations. It should be noted, however, that these general observations are less significant for engineering specialties which tend to have heavier schedules and are more rigidly laid out.

The differences between universities and four-year colleges are less important for lower division work than those between four- and two-year colleges. While the universities permit some slightly greater degree of flexibility in the lower division programs, the more significant differences between universities and four-year colleges relate to upper division work. Although the baccalaureate degrees offered are the same, the variety of specialties available as major fields is much greater in the university, permitting more (or narrower) specialization. This greater number of recognized major fields is reflected in a greater variety of course offerings and in acceptable combinations of courses for meeting graduation requirements in the major field.

In detailed comparisons of physics curricula as offered in Soviet universities and in major U.S. universities (M.I.T. and Columbia) in the mid-1950s, the authors concluded that the science content of the Soviet 5-year physics curricula examined compared favorably with those offered by the best U.S. universities.[4] They found that the Soviet physics curricula were well structured, building on a broad theoretical foundation in the earlier years of study to a highly sophisticated level of specialized training in the last year of study that is not usually included in U.S. education prior to the first or second year of the graduate school program. These conclusions were reaffirmed in a major

[4]Alexander G. Korol, *Soviet Education for Science and Technology*, John Wiley and Sons, p. 357 (1957); and E.M. Corson, *An Analysis of the Five-Year Physics Program at Moscow State University*, U.S. Department of Health, Education and Welfare, p. 1 (1959).

U.S. study of Soviet education made in 1961.[5] In that study, the author found that, although the general physics curricula of Soviet universities had been restructured since the mid-1950s so as to reduce the number of weeks of instruction and intensify the number of weeks of industrial practice assignments, the academic subject matter content had not been significantly affected by these alterations.

Although comparisons of U.S. and Soviet physics curricula indicated that a specialist who had completed the Soviet 5-year physics curriculum (in the regular day program, at least) had a comparable, perhaps somewhat better, professional preparation than a student who had completed four years of college work plus one year of graduate training in physics at a U.S. university, results of studies comparing Soviet to U.S. engineering programs were less conclusive. One author found that "Discounting the vocationally oriented store of precarious knowledge a Soviet engineering graduate possesses after his five years of training, we conclude that in terms of basic engineering preparation he does not achieve appreciably if at all a higher level of competency in his five years of training than his American counterpart does after a 4-year course."[6] Another evaluation conducted about the same time, however, which focused on electrical engineering curricula, found that the level of competency attained by a Soviet student after five-and-one-half years of study corresponded to that of a student who had completed a master's degree at a U.S. university.[7]

Unfortunately, more recent detailed comparisons of curricula in Soviet and U.S. higher educational programs are not available; however, the structure of the Soviet higher educational program does not appear to have undergone any major changes that would call into question the conclusions of these earlier studies.[8] The substantive variations in Soviet undergraduate training from field to field as well as

[5]Nicholas DeWitt, *Education and Professional Employment in the USSR*, National Science Foundation, p. 280 (1961).
[6]Korol, op. cit., p. 357.
[7]P.S. Abetti and G.F. Lincks, *Electrical Engineering Education in the USSR*, as cited by DeWitt, op. cit., p. 785.
[8]An exchange of detailed curricular materials for physics and chemical engineering was included in the original work plans for the Training and Utilization Subgroup of the U.S.-U.S.S.R. Joint Working Group in the Field of Science Policy. Although the U.S. members of the subgroup provided such materials to the Soviet Union, the Soviet side provided such curricular material for only the social sciences and humanities. Search of available literature has not disclosed a current Soviet academic program in the physical sciences or engineering specialties. Thus, while the Soviet members of the Subgroup, to whom detailed curricular materials for both countries were presumably available, stated at a joint meeting in 1978 that in their view the requirements for obtaining a master's degree in the U.S. basically correspond to the requirements for completing an undergraduate education in the USSR, the Soviet curricular materials necessary to confirm this more recent comparison are not available to the authors of this study.

fundamental differences in orientation from U.S. undergraduate programs make direct overall comparisons of the quality of U.S. and Soviet higher education unfeasible. Comparisons of the level of training received in specific specialty areas, however, indicate that the level of professional attainment of a science or engineering graduate of a Soviet higher educational establishment (in the full-time programs, at least) is about the same as, or occasionally higher than, the level of attainment of a science or engineering graduate of a U.S. college or university.

It should be noted, however, that part-time higher educational programs in the Soviet Union, which are generally conceded to offer abbreviated, substandard instruction, still account for approximately 40 percent of total graduation from Soviet higher educational establishments. In addition, while the high degree of specialization and applied functional orientation of the Soviet higher educational process may be an asset in the development of specialists with the ability to attain the short-term technological targets of the Soviet economic plan, the more flexible, broader-based education received in U.S. higher educational institutions appears to produce specialists who are better prepared to meet the longer-term goals of a society with an ability to innovate, an adaptability to technological change, and a greater latitude for interfield mobility as the demands of the economy change.

E. Participation in R&D by Undergraduate Students

In the United States, there is relatively little participation by undergraduate students in research conducted by faculty of higher educational institutions, except for menial laboratory tasks or bibliographical or editing functions that a student might perform under the close direction and supervision of faculty members as part of his responsibilities in connection with a scholarship, work-study program, or other form of financial assistance. In the Soviet Union, by contrast, student participation in research, even at the undergraduate level, is regarded as an important element in the training process. Various methods have been introduced into the higher educational program to encourage independent work by undergraduate students and to foster student interest in scientific research. The type and extent of student participation in research conducted by higher educational institutions vary from one institution to another and even among departments within a given institution. However, in general, students are taught the skills required for bibliographic work, data collection, conducting and analyzing experiments, formulating conclusions, and so forth; and in the later years of the educational program, students often assist with the actual scientific research conducted by the professional teaching staffs of higher educational institutions.

In the earlier years of study in the USSR, organized training is provided for undergraduate students in shops, clinics, experimental farms, and other auxiliary training institutions of the higher educational organization. In seminars in junior courses, students are assigned a problematic theme for independent research. Closely related to seminar projects are courses in which the student receives a project theme to work on, which may eventually become the topic for the student's undergraduate diploma project. In senior courses, students work on research projects related to practical problems. The length and content of the practical work are determined by the academic plans and programs.

In a number of higher educational institutions in the Soviet Union, the formal instruction program is combined with practical research experience, with students spending a period of time during their final year working in branch scientific research institutes and design bureaus under the guidance of regular professional staff members of these institutes. In such cases, the diploma project conducted by the student represents a component of the plan for scientific research work of the research institute or design bureau at which the practical training is conducted.

Apart from the various forms of research experience that are included in the formal academic programs of higher educational institutions in the USSR, there are a number of voluntary programs in Soviet higher educational institutions aimed at expanding student participation in scientific research activity. For example, student scientific circles and design bureaus have been organized at many higher educational institutions in the USSR to promote independent research by students. Student design bureaus generally conduct research at the request of a department laboratory of the higher educational institution with which they are associated or at the request of an industrial enterprise. Projects completed by student design bureaus are sometimes accepted as regular project or course work by the faculty of the higher educational institution, and occasionally even accepted in lieu of an undergraduate diploma project.[9]

To encourage student research, a number of olympiads, competitions, and exhibitions of student scientific work are held in the Soviet

[9]E. Zaleski et al., *Science Policy in the USSR*, Organization for Economic Co-operation and Development, pp. 349–350 (1970).

Union on a regional, city, or departmental level. In addition, All-Union competition for the best work is held annually, with special medals and letters awarded for superior research work.

A major study of Soviet science policy prepared in 1970 for the Organization for Economic Co-operation and Development found a strong participation by Soviet undergraduate students in faculty research. The following examples from the Soviet press cited in that study provide a good illustration of the extent of undergraduate student involvement in research in Soviet higher educational institutions:

> In the scientific research institute for Solid State Physics within the Tomsk Polytechnical Institute, for instance, each student is either attached to an engineer or *aspirant* and has his own individual study plan. The students are familiarized with research during the first or second year of study. They visit laboratories, attend lectures by outstanding scientists, participate in research seminars and regular research conferences and execute concrete research tasks. At the Biologico-soil facility of Leningrad University, students are attached, beginning with their first or second year of study, to those assistants preparing a diploma. During the second year of study the students must present to the Circle an account of their studies. During the third or fourth years, they become laboratory assistants with their study directors, with whom they work in close collaboration. Students' diplomas are prepared during the fifth year of study, on the basis of the preceding work.[10]

Soviet undergraduate student participation in research and development has been estimated as ranging between 15 and 40 percent of total undergraduate enrollment in full-time study.[11] However, adjustments of this in terms of full-time equivalent scientists and engineers engaged in R&D are not available.

F. Enrollment and Graduation

Figure IV-1 shows undergraduate enrollment at higher educational institutions for the United States and the Soviet Union from 1960 to 1978. Figure IV-2 shows USSR diplomas and U.S. bachelor's degrees conferred for the same period. From 1960 to 1978, both total enrollment and total graduation from higher educational institutions in the United States showed a greater increase than in the USSR. In 1978,

---

[10]Ibid., p. 346.
[11]Louvan E. Nolting and Murray Feshbach, "R&D Employment in the USSR—Definitions, Statistics and Comparisons," in *Soviet Economy in a Time of Change*, Vol. I, Joint Economic Committee, U.S. Congress, pp. 735–736 (October 1979).

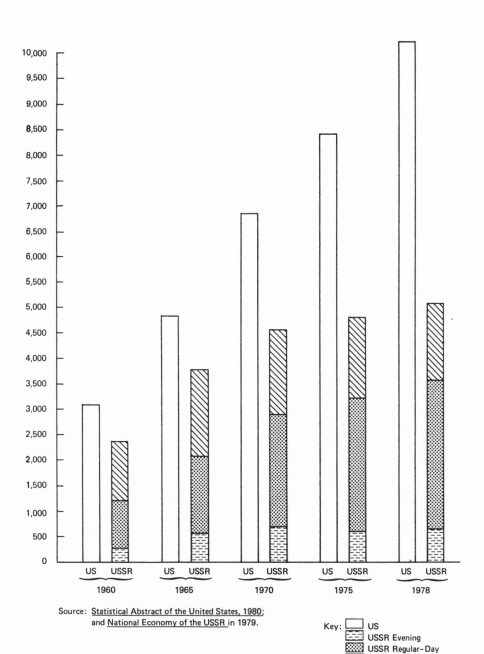

Key: ☐ US
     ▦ USSR Evening
     ▨ USSR Regular–Day
     ◩ USSR Extension
        Correspondence

Figure IV-1   US AND USSR UNDERGRADUATE ENROLLMENT IN HIGHER
EDUCATIONAL INSTITUTIONS: 1960-1978 (In Thousands)

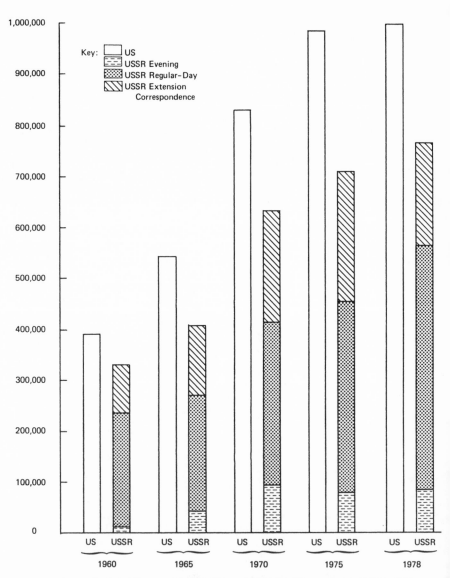

Source: Statistical Abstract of the United States 1980 National Economy of the USSR in 1979.

Figure IV-2 USSR DIPLOMAS AND US BACHELOR'S DEGREES CONFERRED: 1960 - 1978

there were over 10 million students enrolled in undergraduate programs in U.S. higher educational institutions and nearly one million bachelor's degrees conferred. In the Soviet Union during the same year, there were about 5 million students enrolled in higher educational establishments and about 770,000 graduations, with part-time programs accounting for 38 percent of total graduation. As a percent of college age population, the United States in 1978 had about three times the number of students enrolled in higher educational institutions as the Soviet Union had and had about one-and-one-half times as many graduations of the 22/23-year-old population. (See Tables IV-7 and IV-8.)

Table IV-9 shows U.S. bachelor's degrees and Soviet diplomas conferred by major field of study from 1960 to 1979. Data for 1978 are summarized in Figure IV-3. Figures for Soviet graduates in the physical and life sciences and mathematics are estimates based on two-thirds of graduation in "university specialties" to which graduation in geology–prospecting, geodesy–cartography, and hydrology–meteorology have been added.[12] For all fields combined, there were almost 30 percent more graduations from undergraduate programs in the United States than in the Soviet Union. Within the science and engineering fields, the United States graduated about twice as many specialists in the physical and life sciences and mathematics as did the Soviet Union. However, in engineering alone, the Soviet Union graduated almost 6 times the number of specialists graduated in the United States. Even allowing for the probably inferior instruction received by approximately one-third of Soviet engineering graduates who were enrolled in part-time programs (see Section C), the difference is substantial. Largely because of the great number of engineering graduates, in the science and engineering fields combined, the Soviet Union graduated about twice as many specialists as did the United States (8.6 percent of the 22/23-year-old population as opposed to 4.3 percent).

---

[12]Apart from the specialties in geology–prospecting, geodesy–cartography, and hydrology–meteorology, which are classified as engineering by Soviet definitions but which are classified under physical sciences by U.S. definitions, Soviet enrollment and graduation in specialties in the physical and life sciences and mathematics are included in two other of the 22 specialty groups: a large number of specialties in these fields are included in the "university specialties" group and a few are included in the "specialties of pedagogical and cultural higher educational institutions" group. While data are available on enrollment and graduation by specialty group, they are not available by individual specialty. As a rough estimate of enrollment and graduation in the physical and life sciences and mathematics, two-thirds of the total enrollment and graduation in "university specialties" plus that in the geology–prospecting, geodesy–cartography, and hydrology–meteorology groups has been used in this study.

TABLE IV-7

U.S. AND U.S.S.R. UNDERGRADUATE ENROLLMENT IN HIGHER EDUCATIONAL
INSTITUTIONS, TOTAL AND AS A PERCENT OF 18-21/22 YEAR OLDS:[1] 1960-1979

| | U.S. | U.S.S.R. | | | |
	All Institutions[2]	All Institutions	Regular Day	Evening	Correspondence
1960					
Total in Thousands	3,227	2,396	1,156	245	995
Percent of 18-21/22 Year Olds	35.0	11.6	–	–	–
1965					
Total in Thousands	4,829	3,861	1,584	569	1,708
Percent of 18-21/22 Year Olds	39.6	31.2	–	–	–
1970					
Total in Thousands	6,889	4,581	2,241	658	1,682
Percent of 18-21/22 Year Olds	48.7	22.2	–	–	–
1971					
Total in Thousands	7,104	4,597	2,309	647	1,641
Percent of 18-21/22 Year Olds	47.3	21.6	–	–	–
1972					
Total in Thousands	7,199	4,630	2,386	636	1,608
Percent of 18-21/22 Year Olds	46.6	21.4	–	–	–
1973					
Total in Thousands	7,395	4,671	2,463	627	1,581
Percent of 18-21/22 Year Olds	46.8	21.0	–	–	–
1974					
Total in Thousands	7,833	4,751	2,538	632	1,581
Percent of 18-21/22 Year Olds	49.2	20.8	–	–	–
1975					
Total in Thousands	8,468	4,854	2,628	644	1,582
Percent of 18-21/22 Year Olds	51.9	20.8	–	–	–
1976					
Total in Thousands	9,927	4,950	2,711	650	1,589
Percent of 18-21/22 Year Olds	59.7	20.6	–	–	–
1977					
Total in Thousands	10,201	5,037	2,789	652	1,596
Percent of 18-21/22 Year Olds	60.7	20.4	–	–	–
1978					
Total in Thousands	10,179	5,110	2,861	653	1,596
Percent of 18-21/22 Year Olds	60.1	20.4	–	–	–
1979					
Total in Thousands	NA	5,186	2,932	653	1,601
Percent of 18-21/22 Year Olds		20.4	–	–	–

[1] Percent of 18-21 year olds for the U.S. where higher education is generally four years from entrance to graduation, and percent of 18-22 year olds for the U.S.S.R. where higher education is generally around five years from entrance to completion.

[2] Includes both 4-year and 2-year institutions.

Sources: Statistical Abstract of the United States 1978, p. 161;
          Statistical Abstract of the United States 1980, p. 166;
          National Economy of the USSR in 1970, p. 637;
          National Economy of the USSR in 1975, p. 677;
          National Economy of the USSR in 1979, p. 492;
          USSR population data for July 1 of each year, prepared in March 1977 by Foreign Demographics
          Analyses Division, U.S. Bureau of the Census;
          U.S. population data from Population Division, U.S. Bureau of the Census.

TABLE IV-8

U.S. BACHELOR'S DEGREES AND U.S.S.R. DIPLOMAŞ CONFERRED BY HIGHER EDUCATIONAL
INSTITUTIONS, TOTAL AND AS A PERCENT OF 22/23 YEAR OLDS:[1] 1960-1979

	U.S. All Programs	U.S.S.R. All Programs	Regular Day	Evening	Correspondence
1960					
Total in Thousands	395	343	229	15	99
Percent of 22/23 Year Olds	17.6	7.8	-	-	-
1965					
Total in Thousands	539	404	225	44	136
Percent of 22/23 Year Olds	18.1	13.3	-	-	-
1970					
Total in Thousands	833	631	335	82	214
Percent of 22/23 Year Olds	23.9	19.3	-	-	-
1971					
Total in Thousands	884	672	380	86	206
Percent of 22/23 Year Olds	25.3	18.1	-	-	-
1972					
Total in Thousands	938	684	391	91	203
Percent of 22/23 Year Olds	26.9	16.9	-	-	-
1973					
Total in Thousands	980	692	403	90	200
Percent of 22/23 Year Olds	27.1	16.6	-	-	-
1974					
Total in Thousands	1,009	693	417	82	195
Percent of 22/23 Year Olds	27.2	16.2	-	-	-
1975					
Total in Thousands	988	713	433	80	200
Percent of 22/23 Year Olds	26.0	16.4	-	-	-
1976					
Total in Thousands	998	735	448	82	204
Percent of 22/23 Year Olds	25.7	17.1	-	-	-
1977					
Total in Thousands	993	752	462	85	205
Percent of 22/23 Year Olds	25.1	16.8	-	-	-
1978					
Total in Thousands	998	772	479	85	208
Percent of 22/23 Year Olds	24.9	16.2	-	-	-
1979					
Total in Thousands	NA	790	493	85	212
Percent of 22/23 Year Olds	-	16.4	-	-	-

[1] Percent of 22 year olds for the U.S. and of 23 year olds for the U.S.S.R.

Sources: Statistical Abstract of the United States 1978, p. 168;
          Statistical Abstract of the United States 1980, p. 174;
          National Economy of the USSR in 1970, p. 645;
          National Economy of the USSR in 1975, p. 684;
          National Economy of the USSR in 1979, p. 499;
          USSR population data for July 1 of each year prepared in March 1977 by the Foreign Demographics
          Analyses Division, U.S. Bureau of the Census;
          U.S. population data from Population Division, U.S. Bureau of the Census.

TABLE IV-9

U.S. BACHELOR'S DEGREES AND U.S.S.R. DIPLOMAS CONFERRED BY HIGHER EDUCATIONAL INSTITUTIONS, BY MAJOR FIELD OF STUDY: 1960-1979
(In thousands)

Year	1960		1965		1970		1975		1976		1977		1978		1979	
	U.S.	U.S.S.R.	U.S.	U.S.S.R.	U.S.	U.S.S.R.	U.S.	U.S.S.R.	U.S.	U.S.S.R.	U.S.	U.S.S.R.	U.S.	U.S.S.R.	U.S.	U.S.S.R.
All Fields, Total[1]	394.9	343.3	538.9	403.9	833.3	630.8	987.9	713.4	997.5	734.6	993.0	751.9	997.2	771.5	NA	790.0
Science and Engineering, Total	89.4	162.5	109.2	211.8	147.6	328.5	161.4	370.9	164.4	383.4	168.7	397.9	174.0	407.7	NA	416.9
Physical and Life Sciences and Mathematics[2]	45.3	(25.1)	65.7	(25.6)	91.4	(39.7)	96.5	(44.9)	98.2	(46.3)	97.5	(47.8)	95.3	49.0	NA	49.8
Engineering	37.8	102.9	36.8	152.3	44.8	230.5	47.3	272.1	46.7	280.4	49.7	291.4	56.0	300.1	NA	306.8
Agriculture	6.3	34.5	6.8	33.9	11.4	58.3	17.6	53.9	19.5	56.7	21.5	58.7	22.7	58.6	NA	60.3

Percent of 22/23 Year Old Population[3]

Year	1960		1965		1970		1975		1976		1977		1978		1979	
	U.S.	U.S.S.R.	U.S.	U.S.S.R.	U.S.	U.S.S.R.	U.S.	U.S.S.R.	U.S.	U.S.S.R.	U.S.	U.S.S.R.	U.S.	U.S.S.R.	U.S.	U.S.S.R.
All Fields, Total	17.6	7.8	18.1	13.3	23.9	19.3	26.0	16.4	25.6	17.1	25.1	16.8	24.9	16.2	NA	16.4
Science and Engineering, Total	4.0	3.7	3.7	7.0	4.2	10.1	4.3	8.5	4.2	8.9	4.3	8.9	4.3	8.6	NA	8.7
Physical and Life Sciences and Mathematics[2]	2.0	(0.6)	2.2	(0.8)	2.6	(1.2)	2.5	(1.0)	2.5	(1.1)	2.5	(1.1)	2.4	1.0	NA	1.0
Engineering	1.7	2.3	1.2	5.0	1.3	7.1	1.2	6.2	1.2	6.5	1.3	6.5	1.4	6.3	NA	6.4
Agriculture	0.3	0.8	0.2	1.1	0.3	1.8	0.5	1.2	0.5	1.3	0.5	1.3	0.6	1.2	NA	1.3

[1] Includes first professional degrees for U.S.

[2] Figures for the U.S.S.R. are estimates based on two-thirds of graduation in "university specialties" to which graduation in geology-prospecting, geodesy-cartography, and hydrology-meteorology have been added to approximate U.S. definitions.

[3] Percent of 22 year olds for the U.S. and 23 year olds for the U.S.S.R., due to differences in average length of years required to graduate from higher education.

Sources: Earned Degrees Conferred, National Center for Education Statistics, Department of Health, Education and Welfare, Annual Series; National Economy of the USSR in 1970, p. 646 and 1979, p. 500; U.S.S.R. population data from Foreign Demographic Analysis Division, U.S. Bureau of Census; U.S. population data from Population Division, U.S. Bureau of Census.

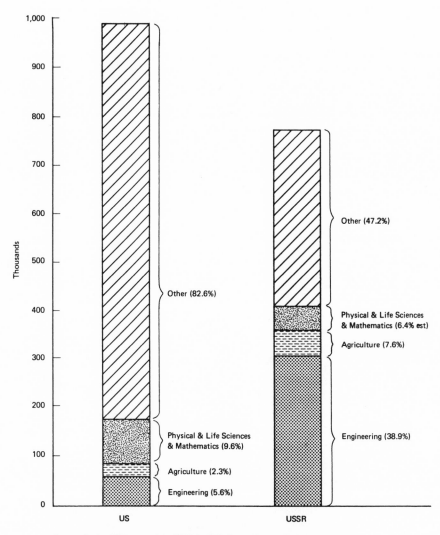

Source: National Economy of the USSR in 1979; Earned Degrees Conferred, 1977-78, Summary Data, National Center for Educational Statistics.

Figure IV-3 U.S. BACHELOR'S DEGREES AND U.S.S.R. DIPLOMAS CONFERRED, BY MAJOR FIELD OF STUDY: 1978

CHAPTER V

Graduate Training

In both the United States and the Soviet Union, two advanced degrees are awarded for training beyond the completion of higher education: the Candidate of Science and the Doctor of Science degrees in the Soviet Union and the Master's Degree and the Doctor of Philosophy degree in the United States. There is no strict comparability, however, between the advanced degrees offered in the two countries. While the awarding of the Candidate of Science degree in the Soviet Union generally involves the completion of an agreed-upon course of study and the preparation and defense of a dissertation, there are exceptions to both the course of study and the dissertation requirement. A large number of students earn the degree by passing examinations and writing and defending a dissertation without ever having been enrolled in a formal program of courses, and the degree is occasionally awarded for the accomplishment of outstanding research work to individuals who have neither taken examinations nor defended a dissertation. The Doctor of Science degree involves no prescribed course of study, but is awarded to more experienced scientists in recognition of outstanding research achievement. In contrast to the Soviet Union, the award of advanced degrees in the United States, although differing in requirements from one university to another, generally involves at both the master's and doctoral level a set sequence of course work, examinations, research, thesis or dissertation, and defense.

In comparing the advanced degrees of the two countries, it is generally believed that there is no equivalent to the U.S. master's degree in the Soviet Union and that the Soviet Candidate of Science degree more or less approximates the U.S. doctoral degree, but at perhaps a slightly lower level of preparation. There are no advanced degrees in the United States comparable to the Soviet Doctor of Science degree, which is essentially an honorary degree awarded to senior research personnel for scientific accomplishments.

Apart from differences in requirements for award of advanced degrees, there are several important aspects of Soviet graduate training which contrast to those in the United States. Advanced degrees in the Soviet Union may be earned not only through training conducted at universities and other institutions of higher education, but also in many of the research organizations under the Soviet academies of sciences and various industrial ministries. Soviet universities, considered by both Soviet and foreign scholars to provide the highest quality training, enroll only about 26 percent of the persons pursuing advanced training, while the majority are trained at research institutions. It appears that the research institute-trained students are more oriented toward meeting the needs of the specific institutions at which they study than toward broader training with more of a theoretical and basic science content. This becomes apparent in the dissertation topics selected and approved by the research institutions as opposed to those dissertations supervised at universities. In addition to lectures, Soviet graduate training includes a significant amount of laboratory research related to the current work of the research institution or the contract work currently in progress at the universities. Thus, Soviet formal graduate training programs tend to have more applied science content than do U.S. programs, especially at the U.S. doctoral degree level.

A. Graduate Training in the USSR

Soviet graduate training is conducted at universities, other institutions of higher education, and many of the research organizations maintained by Soviet academies of sciences and various industrial ministries. Graduate training is offered by over 2,000 such scientific and higher educational institutions. In 1975, graduate training at universities accounted for 26 percent of the training of graduate students.

Two graduate degrees are awarded in the Soviet educational system: Candidate of Science and Doctor of Science. The awarding of the Candidate of Science degree generally occurs after the graduate student has completed an agreed-upon course of study and has written and defended a dissertation. There are exceptions, however, to both the course of study and the dissertation requirements. The Doctor of Science degree involves no prescribed course of study and is generally awarded in recognition of scientific accomplishment rather than for completion of a prescribed educational program.

Entrance into formal graduate study (*aspirantura*) in preparation for the Candidate of Science degree in the Soviet Union is competitive and based on the results of entrance examinations. Some specialists within the system may seek a Candidate degree without formal instruction by basing their research and dissertation preparation on their professional experience. However, more than 70 percent of those seeking Candidate of Science degrees enter graduate programs as aspirants of the degree.

Graduate education in the USSR may be pursued while the student is working or the student may be given time off from employment. More than 62 percent of graduate students in scientific research institutes and about 41 percent of graduate students in universities receive instruction without time off from employment. Full-time graduate study takes three years and must begin before the student is 35 years old. Four years of study are required for those aspirants who do not have time off from work and the person must enter before age 45. There is no age limit for persons who intend to write and defend a dissertation, but who do not enter formal graduate programs as aspirants.

For all graduate-level students pursuing the Candidate of Science degree, whether through formal graduate instruction or independently, an individual plan of study is set up by the student and his academic committee. The general plan of study is confirmed by the academic council of the higher educational institution or research organization, and consists of: preparation for and passing Candidate examinations in dialectical and historical materialism, a special discipline, and a foreign language; and work on the dissertation, including theoretical work, experimental work, and writing the dissertation. A special course of lectures and special seminars on questions and issues involved in the student's area of research is organized for each graduate student. Resident graduate students take part in the academic and research work conducted by the department in which they are studying, and are involved in administering tests and examinations of students, leading seminars, and participating in contract research work. At the end of each year of instruction, the graduate student must pass an examination given by the scientific advisor in the section or department where the student is studying.

Each graduate student is assigned a scientific advisor, who assists the student in selecting his dissertation topic. Full-time graduate students are supposed to do research on a "theoretical level" in an area or field recognized as being on the "frontier of knowledge." Correspondence students generally select a topic related to their jobs and defend their dissertations before the academic councils at the institutions at which they study. Those individuals who are not in formal graduate study but are pursuing a Candidate degree by writing and defending a dissertation also receive assistance from scientific advisors.

The theme of their dissertations must be in accordance with the plan of their specific scientific institution and be confirmed by the academic council.

The area of "special-purpose study" has received greater emphasis in recent years. In this form of study, students are sent to the largest institutions of higher education when they are working in an area that cannot be taught in the smaller universities or other institutions in which they are enrolled. Eighty percent of this type of study is conducted in higher educational institutions. After completion of their studies, special-purpose students return to their own organizations. Between 1971 and 1973, 36 percent of all individuals entering graduate study were in special-purpose programs. About half of the full-time graduate students at Moscow State University are special-purpose students.

Students undertaking graduate level training are generally supported by either special state stipends or by the organizations for which they work. Full-time graduate students in scientific research organizations and higher educational institutions are paid stipends. For graduate students who enter graduate school with less than 2 years of employment experience, stipends are paid in the amount of the basic official salary scale received before enrollment in graduate school, but not higher than 100 rubles a month. Stipends are not paid to students enrolled in graduate school who do not take time off from employment; these students are given supplementary monthly leave with pay and a day off each week at half pay. During the total period of their training, part-time graduate students are entitled to up to six months of leave. Graduate students are also given a number of special privileges: transportation at a privileged tariff (from October to May), summer vacation (for those studying with time off from employment), additional leave (for correspondence graduate students), free use of the library, etc. Those studying out of town are provided housing.

B. Graduate Training in the United States

In contrast to Soviet graduate training, graduate degrees in the United States can be earned only through formal programs at institutions of higher education. The master's degree program is generally from one to two years in duration, while the doctorate requires three to six years, depending on the field and institution. Specific requirements for graduate level degrees in the United States vary greatly from university to university and from department to department. The number as well as variety of course options, thesis proposals, written and oral preliminary

and comprehensive examinations, language and research tool requirements, and academic standards is large. However, a number of general
observations can be made.[1]

Most U.S. master's degree programs require the equivalent of two
semesters of full-time study after completion of the bachelor's degree.
In the case of a master's degree for some professions (i.e., a Master
of Business Administration), however, study can extend to two or
more full academic years. For degrees in fields requiring a large amount
of data work for a thesis, the program often includes a full year of
course work and an additional year for thesis research. There is generally a time limit, usually about five years, within which all requirements for the degree must be completed.

Most U.S. graduate schools specify some combination of thesis and/
or written or oral comprehensive examination to fulfill the requirements
for the master's degree. A number of programs have two plans for the
master's degree: one which includes a thesis or final project and one
in which the student elects to complete additional course work in lieu
of a thesis. The master's thesis may take on a variety of forms. It may
be a long research paper, an essay or a short report of a project, a
seminar paper, or any combination of the above. A final comprehensive
examination is given in almost all cases, although the form and content
of the examination vary greatly.

The Doctor of Philosophy degree (Ph.D.) is the highest degree
granted by a university in the United States. It represents superior
attainment in one field and is a demonstration of the ability to conduct
independent research and contribute to the body of knowledge in that
area. The amount of time required to complete the course work for the
Doctor of Philosophy degree varies greatly from school to school, but
is usually about three years of full-time study or its equivalent beyond
the bachelor's degree. Time in which the dissertation is to be written
is usually allotted following this period. The time limit for completion
of the dissertation and all other requirements for the Ph.D. is usually
about seven years. Ph.D. programs in the United States require a thesis
or dissertation. The doctoral dissertation is expected to be the result
of exhaustive research into one specific field of interest which provides

[1]In order to obtain a picture of graduate programs in the United States, the graduate school
catalogs from a sample of 31 universities were reviewed during preparation of the U.S. survey
report ("The Training and Utilization of Scientific and Engineering-Technical Personnel," SRI
International, SSC-TN-4226-1, 1979). The sample schools were chosen by geographical location, control of institution (public and private), and enrollment size. From these catalogs, a
general picture of time, language, thesis, examination, and grade point requirements was obtained, showing both the similarities among programs in some areas and the wide variation in
others.

an original contribution to the body of knowledge in that area or which presents a new and significant interpretation. Most graduate programs require an oral defense of the dissertation following its completion.

Graduate student participation in research work in the United States varies for different institutions and different fields, with the highest participation in research activities occurring in the science fields. Many graduate programs have a requirement that students take courses in application of research methodology, in which the student may perform certain laboratory experiments or gather or organize research materials. In addition, there is a form of financial aid for graduate students, known as a research assistantship, which may require that as much as one-half of the student's time be devoted to carrying out research responsibilities, normally under the close direction and supervision of a faculty member.

Candidates for the doctoral degree generally conduct at least some research work in the preparation of their dissertations. The amount of research and the nature of the work involved vary for different fields, different programs, and different institutions. Research conducted in the preparation of a dissertation is generally designed and implemented by the student independently, although he may receive guidance from his faculty dissertation advisor.

While the doctoral degree is the highest formal degree awarded in U.S. higher educational programs, since World War II there has been an increasing trend among doctoral degree recipients to extend their education beyond that of the doctoral degree program by obtaining temporary appointments to research positions at universities, government laboratories, research institutions, or industrial laboratories.[2] A number of factors have contributed to the popularity of postdoctorals in the United States. The rapidly accelerating growth of science and its continuing fragmentation into new fields of specialization have made it desirable for many students to try to remain in the academic environment for one or more years after receipt of the doctorate before beginning permanent, professional employment. The postwar availability of federal research grants and other sources of support created many opportunities for Ph.D.s to obtain temporary appointment as research associates or as postdoctoral fellows. Shortage of openings in some fields on the faculties of colleges and universities or in industry has further motivated students to seek such temporary positions. Fi-

---

[2]See *The Invisible University, Postdoctoral Education in the United States,* report of a study conducted under the auspices of the National Research Council, National Academy of Sciences (1969).

nally, college science departments have found it convenient to create such positions as a holding pattern to retain exceptionally gifted Ph.D.s. The creation of such positions serves to strengthen the science departments at relatively low expense and provides an opportunity for the appointees to teach advanced courses and engage in independent specialized research.

C. Enrollment and Graduation of Graduate Students

Soviet data on enrollment and graduation of graduate students show only those students who are enrolled in formal graduate training (*aspirantura*) leading to the Candidate of Science degree. Thus, not included in the data are those students who are seeking a Candidate degree without entering a formal graduate program, those students who have graduated from a formal graduate program and are currently studying for Candidate examinations or working on their Candidate dissertation or defense, and those students who are conducting their research for the Doctor of Science degree. In addition, while data are published on the total number of Doctor of Science and Candidate of Science degree holders in the economy as a whole, data on the number of advanced degrees conferred annually are not published in the Soviet statistical handbooks.[3] The incompleteness of Soviet published data on the number of students pursuing a graduate education and on advanced degrees conferred makes strictly numerical comparison of U.S. and Soviet graduate education difficult.

Apart from the differences in reported statistics noted above, a numerical comparison of Soviet and U.S. graduate education is further complicated by the fact that the levels of qualification of U.S. and Soviet advanced-degree holders are not strictly comparable: the Candidate of Science degree appears to be a higher level of qualification than the U.S. master's degree but slightly less than the U.S. doctoral degree; the Soviet Doctor of Science degree appears to be awarded to persons of greater expertise, particularly in terms of research accomplishments, than the U.S. doctoral degree. Thus, direct numerical comparisons are more useful in illustrating relative trends in graduate education in the two countries than as indicators of net additions to the stock of advanced-degree holders.

Relative trends in enrollment and graduation in graduate education

[3] Efforts have been made to estimate USSR advanced degrees conferred annually by field of science; see, for example, Joseph S. Conlin Jr., *Soviet Professional Scientific and Technical Manpower*, Defense Intelligence Agency, ST-CS-01-49-74 (October 1973).

in the United States and the Soviet Union are shown in Figures V-1
and V-2. The USSR shows a distinct leveling off in both enrollment
and graduation of aspirants beginning in 1967-68, with an actual decline in the most recent years. The United States showed a consistent
increase in enrollment of graduate students through 1975, but a decrease of over 14 percent from 1975 to 1978. U.S. graduation of
graduate students continued to increase through 1977, but began to
decline in 1978.

Table V-1 shows the breakdown of graduate student enrollment in
the United States and the Soviet Union by major field of study from
1960 to 1974, the latest year for which such a breakdown is available
for the Soviet Union. About 75 percent of Soviet aspirant enrollment
was in the fields of science and engineering as compared with about
20 percent of total U.S. master's and doctoral enrollment occurring
in these fields. It should also be noted that the percentage of total U.S.
graduate enrollment in these fields has been steadily declining since
1960, when it accounted for about 30 percent of the total.[4] While the
data in the table show total U.S. graduate enrollment in science and
engineering as consistently much higher than the corresponding Soviet
enrollment, the reader should be reminded once again that as the Soviet
data show only those students who are enrolled in formal aspirant
training and do not include students who are preparing for examinations
or working on their dissertation or defense, these data should be viewed
only as illustrating relative trends in the two countries.

It should be noted that the United States has a far greater proportion
of graduate students enrolled in the field of education than does the
Soviet Union. This has implications for the quality of general training
of students in the two educational systems although this aspect of
education is difficult to quantify. To the extent that an advanced degree
denotes a higher quality input into the educational system, it would
appear that the United States is staffing its schools, grades K-12 and
colleges, with persons with more graduate training, and thus of higher
quality, than is the Soviet Union.

Table V-2 shows the number of U.S. and Soviet specialists with
advanced degrees (Soviet Candidate of Science and Doctor of Science
degrees and U.S. doctoral degrees, which represent the nearest U.S.
equivalent) by field. As a quantitative indicator of the magnitude of
graduate education in the two countries, these data provide a more

[1]Data for 1976 further confirm this trend. In 1976, 14.1 percent of U.S. graduate enrollment
was in the physical and life sciences and mathematics and 5.6 percent was in engineering for
a total of 19.7 percent in the science and engineering fields. (*Digest of Education Statistics
1980*, p. 95.)

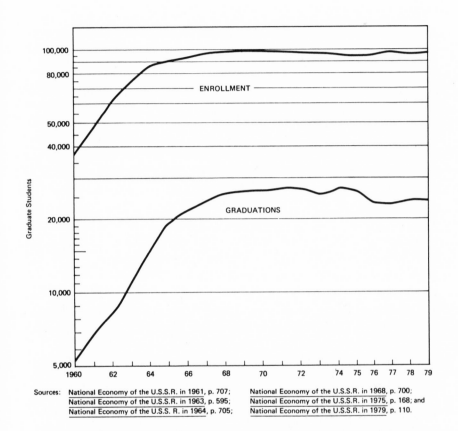

Sources: National Economy of the U.S.S.R. in 1961, p. 707;   National Economy of the U.S.S.R. in 1968, p. 700;
          National Economy of the U.S.S.R. in 1963, p. 595;   National Economy of the U.S.S.R. in 1975, p. 168; and
          National Economy of the U.S.S. R. in 1964, p. 705;   National Economy of the U.S.S.R. in 1979, p. 110.

Figure V-1   U.S.S.R. ENROLLMENT AND GRADUATIONS IN GRADUATE EDUCATION (ASPIRANTS): 1960 to 1979

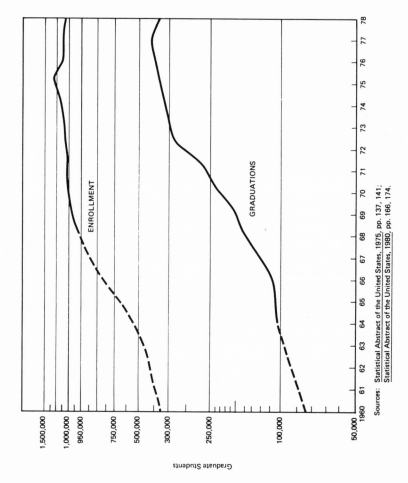

Graduate Students

Sources: Statistical Abstract of the United States, 1975, pp. 137, 141; Statistical Abstract of the United States, 1980, pp. 166, 174.

Figure V-2   U.S. ENROLLMENT AND GRADUATIONS IN GRADUATE EDUCATION: 1960 to 1978

TABLE V-1

U.S. AND U.S.S.R. GRADUATE STUDENT ENROLLMENT,
BY MAJOR FIELD OF STUDY: 1960-1974

	1960				1965				1970				1974			
	U.S.		U.S.S.R.		U.S.		U.S.S.R.		U.S.		U.S.S.R.		U.S.		U.S.S.R.	
	Number	Percent	Number	Percent	Number	Percent	Number	Percent	Number	Percent	Number	Percent	Number	Percent	Number	Percent
Physical & Life Sciences and Mathematics[1]	58,094	18.5	11,612	31.6	93,594	17.5	29,537	32.7	121,463	14.9	30,057	30.2	137,950	14.3	27,138	28.0
Engineering	36,636	11.7	13,936	37.9	57,516	10.7	35,733	39.6	64,788	8.0	39,979	40.2	56,037	5.8	39,977	41.2
Agriculture	5,493	1.7	2,877	7.8	8,039	1.5	7,323	8.1	10,432	1.3	6,312	6.3	12,601	1.3	5,449	5.6
Education	94,993	30.2	956	2.6	150,300	28.1	1,480	1.6	257,605	31.6	2,097	2.1	327,113	33.9	2,242	2.3
Other	119,133	37.9	7,373	20.1	225,883	42.2	16,221	18.0	361,919	44.3	20,982	21.1	431,299	44.7	22,163	22.9
Total	314,349	100.0	36,754	100.0	535,332	100.0	90,294	100.0	816,207	100.0	99,427	100.0	965,000	100.0	96,939	100.0

[1] Includes both masters and doctorate degree enrollment in biological sciences, computer and information sciences, health professions, mathematics and physical sciences for the U.S.; includes enrollment in formal aspirant training in biology, chemistry, geology and mineralogy, medicine and pharmacy, and physics and mathematics for the U.S.S.R.

Source: Digest of Education Statistics 1979, p. 96, National Center for Education Statistics;
   The National Economy of the U.S.S.R. in 1960, p. 789;
   The National Economy of the U.S.S.R. in 1965, p. 716;
   The National Economy of the U.S.S.R. in 1970, p. 662;
   The National Economy of the U.S.S.R. in 1974, p. 148.

TABLE V-2

U.S. AND U.S.S.R SPECIALISTS WITH ADVANCED DEGREES,[1]
BY BRANCH OF SCIENCE AND ENGINEERING: 1974-1977
(Actual Figures for United States;
Soviet Figures Rounded to Nearest Hundred)

Branches of Science in which employed	U.S. beginning of 1977		U.S. beginning of 1975	U.S.S.R., end of 1974	
	Number	Per 1000 in Labor Force	Number	Number	Per 1000 in Labor Force
Physical and Life Sciences	147,607	148.3	143,976	160,000	160.3
Physics/Mathematics	34,778	34.9	34,631	36,500	36.5
Physics and Astronomy	17,911	17.9	16,793	–	–
Mathematics	16,867	16.9	17,838	–	–
Chemistry	29,640	29.7	29,548	19,500	19.5
Environmental Sciences	14,170	14.1	13,842	15,200	15.2
Geology/Mineralogy	4,081	4.1	4,023	11,600	11.6
Geography	7,103	7.1	6,140	3,600	3.6
Other Earth Sciences	2,986	2.9	3,679	–	–
Biology	39,324	39.4	38,463	25,300	25.3
Agriculture	10,641	10.6	10,070	17,100	17.1
Medicine/Pharmaceutics	18,164	18.1	16,678	43,000	43.0
Veterinary Sciences	890	0.8	744	3,400	3.4
Engineering (Technical Sciences)	49,481	49.7	44,113	102,900	103.1
Social Sciences	72,526	72.8	–	34,400	34.4
Economics	16,376	16.0	14,408	23,300	23.3
Law (Political Science)	17,735	17.7	–	3,600	3.6
Education	10,467	10.5	–	6,200	6.2
Psychology	27,948	28.0	27,596	1,300	1.3
Humanities	55,355	55.6	–	35,700	35.7
Philosophy/Sociology	10,070	10.1	–	8,000	8.0
Language and Literature	26,040	26.1	–	12,600	12.6
History	14,237	14.3	–	13,400	13.4
Art	5,008	5.0	–	1,700	1.7
Architecture	923	0.9	–	1,100	1.1
Other (U.S.)	13,275	13.8	–	–	–
Other (U.S.S.R.)	–	–	–	7,100	7.1
Total	339,167	340.9	–	341,200	341.8

[1] For U.S., "Advanced Degrees" includes doctorates, for U.S.S.R., they
include both Doctors of Science and Candidates of Science.

Sources: Louvan E. Nolting and Murray Feshbach, "R&D Employment in the USSR-
Definitions, Statistics and Comparisons" in <u>Soviet Economy in a
Time of Change</u>, Vol. I, Joint Economic Committee, U.S. Congress,
p. 750 (October 1979); and labor force figures from <u>The National
Economy of the U.S.S.R. in 1974</u>, p. 549; and <u>Statistical Abstract
of the United States, 1978</u>, p. 398.

realistic picture than do comparisons of figures on graduate enrollment, for the reasons discussed above. This table shows that the Soviet Union now has approximately the same total number of specialists with advanced degrees as does the United States, and more than twice the number of engineers with advanced degrees. From among the major aggregate branches of science, the United States leads the Soviet Union in the number of specialists with advanced degrees in only the social sciences and the humanities. However, if agriculture and medicine are treated separately from the physical and life sciences, the United States showed an aggregate total for physics/mathematics, chemistry, environmental sciences, and biology of 116,480 at the beginning of 1975 compared to 96,500 for the USSR at the end of 1974.

It is important to also take into consideration, however, the number of U.S. master's degree holders, for which there is no Soviet equivalent. The Soviet aspirant graduate has completed the required instructional components of his graduate training and has only to complete the dissertation to be awarded the Candidate of Science degree; he thus has a higher level of training than a U.S. master's degree holder. Master's degree holders, however, represent a large reservoir of persons with scientific training who fill critical middle-echelon positions in U.S. industry and government. In 1978, there were over 300,000 master's degrees conferred by U.S. universities, with about 40,000 in the science and engineering fields. (See Table V-3.)

Figure V-3 shows the number of U.S. master's and doctoral degrees relative to the estimated number of Soviet Candidate degrees conferred in the engineering fields from 1960 to 1976. When master's degrees are included in assessing the total number of engineers being produced annually with degrees higher than the baccalaureate level, the United States leads the Soviet Union. There are, however, substantially more Candidate degrees conferred annually in the engineering fields than doctorates, which represent the nearest U.S. equivalent.

TABLE V-3

U.S. MASTER's DEGREES CONFERRED BY MAJOR FIELD OF STUDY: 1978

Agriculture & Natural Science	4,036
Architecture, Environmental Design	3,121
Area Studies	925
Biological Sciences	6,851
Business and Management	48,661
Communications	3,297
Computer & Information Sciences	3,038
Education	118,957
Engineering	16,409
Fine & Applied Arts	9,036
Foreign Languages	2,741
Health Professions	14,483
Home Economics	2,613
Law	1,786
Letters	10,062
Library Sciences	6,935
Mathematics	3,383
Military Sciences	45
Physical Sciences	5,576
Psychology	8,194
Public Affairs & Services	20,191
Social Sciences	14,660
Theology	3,329
Interdisciplinary Studies	4,487
Total	312,816

Source: Earned Degrees Conferred, 1977-78, Summary Data,
          National Center for Education Statistics, pp. 28-32.

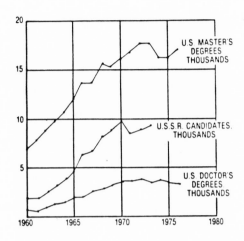

Source: Roger K. Talley, <u>Soviet Professional Scientific and Technical</u>
          <u>Manpower</u>, Defense Intelligence Agency, DST-1830S-049-76,
          (May 1976).

Figure V-3   U.S. AND U.S.S.R. ADVANCED DEGREES CONFERRED IN ENGINEERING:
                                 1960-1976

CHAPTER VI

The Scientist, Engineer, and Technician Populations

Definitions used for classifying individuals as scientists, engineers, or technicians differ with respect to the statistical reports of the United States and the Soviet Union. As a result, difficulties arise in comparing absolute numbers of scientific, engineering, and technical personnel in the two countries. This chapter will describe the definitions used in the two reporting systems, and discuss the data that are available on the scientist, engineer, and technician populations employed in the economies of the Soviet Union and the United States. U.S.-Soviet comparisons of total scientific and engineering manpower will be presented, but more as an indication of relative trends in the two economies than as directly comparable data. Chapter VII will then focus on the utilization of scientists and engineers in the R&D employment sectors of the two countries.

A. *The Technician Populations in the USSR and the United States*

Direct comparison of the number of technicians in the United States and the Soviet Union is complicated by differing definitions of "technician," a situation which arises from the often diverse applications of a technician's training. In the American case, the training itself is often as diverse as its application, while in the Soviet Union, the majority of technicians are educated in specialized secondary educational institutions.

In a response to a U.S. inquiry on the role of technicians in the Soviet economy, Soviet specialists responded as follows:

> Specialized secondary education institutions meet the requirements of the national economy for middle level specialists: foremen, heads of shifts, sections and transportation systems. They also prepare technician-technol-

ogists and technician-designers for the technological and design services of enterprises. Graduates of secondary specialized education institutions resolve particular questions and general engineering and scientific tasks, they prepare documentation, experimental data and other materials in project-design and scientific-research organizations. Graduates of the technical school also perform direct work operations connected with the maintenance of complicated equipment-machines, assemblers, and automatic lines. Specialist-technicians who directly organize production know the concrete production process well and have administrative skills in directing production. The growth of the productivity of labor, the correct use of the achievements of science and technology, the clear-cut organization of labor and production largely depends on their knowledge and experience.[1]

Table VI-1 shows the number of employed graduates of specialized secondary education, by category of specialization, from 1955 to 1975. There has been a steady increase in the number of engineering technicians as a percent of total specialists with specialized secondary education, rising from 27.9 percent in 1955 to 45.7 percent in 1975. The increase in the percentage of specialists with specialized secondary education employed as planners and statisticians has been especially great during the period, rising from 5.2 percent in 1955 to 11.0 percent in 1975.

Table VI-2 shows the ratio of technicians to professionals with higher education in the Soviet Union from 1955 to 1979. Throughout the period, the medical sciences have had the highest number of technicians per professional worker, with a ratio of 2.9:1 in 1977, while the educational and legal fields have had the lowest number of technicians per professional worker, with ratios of 0.5:1 and 0.2:1, respectively, in 1977. Since 1960, the number of technicians per professional worker on an economy-wide basis has been consistently declining, with a ratio of 1.48:1 in 1960 and a ratio of 1.38:1 in 1979. In the industrial sector, where a ratio of 3 to 4 technicians per engineer is considered desirable by Soviet planners, the actual ratio was 1.6:1 in 1977, a decline from the 1965 ratio of 1.8:1.[2] This is considered one of the causes of the widely recognized problem of underutilization of engineers in the Soviet Union, as without adequate technician support, professionals must

[1]Soviet members of the U.S.-USSR Subgroup on Training and Utilization of Scientific, Engineering and Technical Personnel, in "Answers to Questions of the American Experts on the First Part of the Soviet Report on Training and Utilization of Scientific and Engineering-Technical Personnel in the USSR," Moscow (1977).
[2]Jill E. Heuer, "Soviet Professional, Scientific, and Technical Manpower," Defense Intelligence Agency, DST-1830 S-049-79, p. 4 (May 1979).

TABLE VI-1

U.S.S.R. SPECIALISTS WITH SPECIALIZED SECONDARY
EDUCATION OCCUPIED IN THE NATIONAL ECONOMY,
BY SPECIALTY: 1955 to 1975
(In Thousands)

	1955	1960	1965	1970	1973	1975
Total Number of specialists occupied in the economy	2949.1	5238.5	7174.9	9988.1	11976.7	13319.3
By specialty, achieved in school						
Technicians	822.6	1955.8	2886.7	4333.1	5373.3	6092.6
Percent of total	27.9	37.3	40.2	43.4	44.9	45.7
Agronomists, zootechnicians, veterinary assistants, veterinary technicians	236.7	356.3	465.0	597.0	698.2	766.4
Percent of total	8.0	6.8	6.5	6.0	5.8	5.8
Planners and statisticians	152.8	337.5	571.0	950.5	1248.4	1464.9
Percent of total	5.2	6.4	8.0	9.5	10.4	11.0
Commodity specialists	33.3	106.7	249.2	396.3	520.0	605.9
Percent of total	1.1	2.0	3.5	4.0	4.3	4.5
Lawyers	23.2	17.2	16.4	21.1	25.9	29.8
Percent of total	0.8	0.3	0.2	0.2	0.2	0.2
Medical workers (including dentists)	731.1	1187.3	1453.6	1862.1	2124.9	2276.6
Percent of total	24.8	22.7	20.3	18.6	17.7	17.1
Teachers, librarians, cultural- educational workers	818.6	1061.9	1282.3	1458.9	1558.9	1595.4
Percent of total	27.8	20.3	17.9	14.6	13.0	12.0

Note: Numbers reported by specialty add to only about 96 percent of the
total number of specialists reported.

Source: National Education, Science and Culture in the U.S.S.R, p. 293
(1977) and p. 235 (1971).

TABLE VI-2

U.S.S.R. SEMIPROFESSIONAL TECHNICIANS PER PROFESSIONAL,
BY OCCUPATIONAL SECTOR: 1955 to 1979

SECTOR	1955	1960	1965	1970	1973	1975	1977	1979
Engineers and industrial technicians	1.4	1.7	1.8	1.7	1.7	1.7	1.6	
Agricultural specialists	1.6	1.6	1.5	1.5	1.5	1.5	1.5	
Economists/planners/statisticans	1.6	1.7	1.9	1.9	1.9	1.9	1.9	
Teachers and cultural workers	0.9	0.8	0.7	0.6	0.5	0.5	0.5	
Legal specialists	0.5	0.3	0.2	0.2	0.2	0.2	0.2	
Medical specialists	2.4	3.0	2.9	3.1	3.0	3.0	2.9	
Entire economy	1.35	1.48	1.47	1.46	1.43	1.41	1.39	1.38

Source: The National Economy of the USSR in 1960, p. 650.
         The National Economy of the USSR in 1965, p. 574.
         The National Economy of the USSR in 1973, p. 592.
         The National Economy of the USSR in 1977, p. 393.
         The National Economy of the USSR in 1979, p. 397.

spend a considerable portion of their time on administrative and support functions.[3]

As noted in Chapter III, there is a wide diversity in the level of training of persons classified as technicians in the United States. The U.S. data on technicians employed in science and engineering used herein are based on a definition of technicians as persons engaged in technical work at a level requiring knowledge equivalent to training received through two years of post high school education. Persons engaged in such activities are included as technicians even if they hold a bachelor's or higher degree.

Table VI-3 shows employment of technicians in the United States for the period from 1960 to 1970. During the decade, there was a 52 percent increase in the total number of technicians employed, with an average annual rate of growth of 4.3 percent. The growth of technicians in the life sciences was particularly high at 69.2 percent, with an average annual rate of growth of 5.4 percent.

Table VI-4 shows the number of employed technicians in the United States, by occupation and employer, for 1970. In terms of type of employer, 41.3 percent of all employed technicians were in manufacturing, 18.9 percent in government (9.8 percent in federal, 6.5 percent in state, and 2.6 percent in local), 4.3 percent in universities and colleges, and 0.7 percent in nonprofit institutions.

Of the U.S. employed technicians, 36.1 percent were in engineering, 29.6 percent in drafting, 12.0 percent in physical sciences, 8.3 percent in life sciences, and 13.9 percent in other types of fields. Draftsmen are technicians who prepare detailed drawings based on rough sketches, specifications, and calculations of architects, designers, and engineers; thus, it is not surprising that, although the majority worked for manufacturing firms, over one-quarter of them worked for engineering and architectural firms. Engineering technicians work in all phases of production, and most were employed by manufacturing firms. Within manufacturing, they were highly concentrated in electrical equipment. Physical and life sciences technicians are primarily laboratory assistants. The largest block of physical science technicians were in manufacturing, with about one-third of these involved in the manufacture of chemicals. More than one-quarter of the life science technicians worked in medical and dental laboratories, and another one-quarter worked for universities and colleges.

---

[3]See Section D of this chapter for a fuller discussion of this issue.

TABLE VI-3

U.S. EMPLOYMENT OF TECHNICIANS, [1] BY OCCUPATION: 1960 to 1970

	1960	1962	1964	1966	1968	1970	PERCENT GROWTH OVER DECADE
Total	676,400	749,900	828,900	903,600	988,500	1,027,800	52.0%
Draftsmen	205,000	220,600	241,400	273,800	299,400	304,400	48.5%
Engineering	237,900	270,100	300,900	322,600	356,500	371,500	56.2%
Physical Sciences	84,700	93,300	100,600	107,100	116,300	123,000	45.2%
Life Sciences	50,600	57,800	66,200	70,100	78,800	85,600	69.2%
All Other	98,200	108,100	119,800	130,000	137,500	143,300	--

[1] Generally defined here as employed at a level equivalent to training received through two years of post high school education.

Source: Training and Utilization of Scientific and Engineering-Technical Personnel: Appendices, SRI International, pp. 74-79 (November 1976), citing unpublished data from Bureau of Labor Statistics sources.

TABLE VI-4

U.S. EMPLOYED TECHNICIANS,[1] BY OCCUPATION AND TYPE OF EMPLOYER: 1970

Type of Employer	Total	Draftsmen	Engineering	Physical Sciences	Life Sciences	All Other
Employers, total	1,027,800	304,400	371,500	123,000	85,600	143,300
Mining	12,200	3,900	2,800	1,300	200	4,000
Contract construction	36,100	21,900	7,900	200	0	6,100
Manufacturing	424,600	144,400	161,400	63,800	7,200	47,800
Transportation, communications, and public utilities	69,500	10,200	46,700	2,600	200	9,800
Other industries	247,400	101,400	45,400	20,700	29,500	50,400
Medical and dental laboratories	24,900	0	0	0	22,400	2,500
Nonprofit institutions	6,800	700	900	1,500	2,800	900
Engineering/architectural services	115,800	81,200	8,900	1,000	0	24,700
Government	193,900	19,800	96,500	28,300	25,600	23,700
Federal	100,600	3,100	53,600	24,300	14,900	4,700
State	66,300	7,500	35,100	2,600	9,500	11,600
Local	27,000	9,200	7,800	1,400	1,200	7,400
Colleges and universities	44,100	2,800	10,800	6,100	22,900	1,500

[1] Generally defined as employed at a level equivalent to training received through two years of post high school education.

NOTE: Because of rounding, sums of individual items may not add to totals.

Source: Training and Utilization of Scientific and Engineering-Technical Personnel: Appendices, SRI International, p. 79 (November 1976), citing unpublished data from Bureau of Labor Statistics sources.

B. The Scientist and Engineer Population in the USSR

Published data in the Soviet Union on the scientist population are based on the number of "scientific workers." Scientific workers include everyone with an advanced degree (Doctor of Science or Candidate of Science), irrespective of his place or type of work, and everyone conducting scientific research work in scientific establishments or teaching in higher educational institutions. Instructions accompanying the statistical questionnaires used by the Soviet Central Statistical Administration define "scientific workers" as:

Academicians, full members and corresponding members of all academies of science; all persons with an advanced degree of Candidate or Doctor of Science or with the scientific title of professor, associate professor, assistant professor, senior scientific worker, junior scientific worker or academic assistant, irrespective of the place or nature of their work; and also all persons carrying out scientific research work in scientific institutions or carrying out research and teaching in higher educational establishments irrespective of whether they have advanced degrees or scientific titles; and also specialists without an advanced degree or scientific title carrying out research at industrial enterprises and project organizations on a regular basis.[4]

In the Soviet statistics, scientific workers are distributed under the following 20 fields or "branches" of science:

Physics and mathematics
Chemistry
Biology
Geology and mineralogy
Technical sciences
Agriculture
History
Economics
Philosophy
Philology
Geography
Law
Education
Medicine
Pharmacy

[4]*The National Economy of the U.S.S.R. in 1975*, Moscow "Statistika," p. 780.

Veterinary sciences
Art
Architecture
Psychology
Other (military and military-related).[5]

The "technical sciences" branch includes a large number of subdisciplines that would largely be considered engineering fields in the United States.

Table VI-5 shows the distribution of scientific workers by branch of science from 1950 to 1974. Table VI-6 shows the average annual rate of growth of scientific workers over the same period. There was a considerable increase in the number of scientific workers in each branch of science. The rate of increase, however, varied widely between branches, with the technical sciences showing one of the highest rates of increase throughout the period. In 1974, approximately 47 percent of the total number of scientific workers were in the technical sciences, 21 percent in the natural sciences, and 20 percent in the social sciences and the humanities. The branches of science in which there were the highest numbers of scientific workers in 1974 (the technical sciences, physics and mathematics, and economics branches) were also among those branches which showed a higher rate of increase than the number of scientific workers as a whole. During the period from 1950 to 1974, physics and mathematics went from sixth to second highest in terms of number of scientific workers, and economics changed from tenth to third. The decrease between 1970 and 1974 in the number of scientific workers in education has been attributed to the reorganization of several pedagogical higher educational institutions into universities and related changes in the system of accounting for scientific-pedagogical workers in different departments.

Soviet data on engineers report the number of persons employed in the economy who have received diplomas in engineering specialties from higher educational institutions, regardless of the character of the work in which they are actually engaged. The following branches in higher educational institutions are classified as engineering:

Geology and exploration of mineral deposits
Exploitation of mineral deposits (mining)

---

[5]Listed in the 1972 "Nomenclature of Specialties of Scientific Workers," *Decree of Minvuz USSR*, No. 647, August 1972, as military sciences and naval sciences but not listed separately in the 1977 "Nomenclature." See Nolting and Feshbach, op. cit., p. 752, footnote.

TABLE VI-5

U.S.S.R. SCIENTIFIC WORKERS, BY BRANCH OF SCIENCE: 1950 TO 1974

	1950		1955		1960		1965		1970		1974	
	Number	Percent of Total	Number	Percent of Total	Number	Percent of Total	Number	Percent of Total	Number	Percent of Total	Number	Percent of Total
TOTAL	162,508	100.0	223,893	100.0	354,158	100.0	664,584	100.0	927,709	100.0	1,169,700	100.0
Physical and Life Sciences	70,958	43.7	93,635	41.8	134,325	37.9	208,143	31.3	284,174	30.6	341,300	29.2
Physics & mathematics	10,184	6.3	20,077	9.0	28,966	8.2	63,880	9.6	95,272	10.3	116,900	10.0
Chemistry	12,946	8.0	16,435	7.3	26,237	7.4	33,534	5.0	45,815	4.9	53,700	4.6
Biology	8,621	5.3	11,009	4.9	15,091	4.3	27,057	4.1	37,342	4.0	45,500	3.9
Geology/mineralogy	3,626	2.2	5,653	2.5	10,671	3.0	16,441	2.5	20,342	2.2	24,500	2.1
Medicine	21,040	12.9	24,807	11.1	31,393	8.9	35,752	5.4	48,750	5.3	57,600	4.9
Pharmacy	435	0.3	519	0.2	781	0.2	908	0.1	1,207	0.1	1,400	0.1
Agriculture	11,932	7.3	12,801	5.7	17,970	5.1	27,076	4.1	31,146	3.4	36,500	3.1
Veterinary sciences	2,174	1.3	2,334	1.0	3,216	0.9	3,495	0.5	4,300	0.5	5,200	0.4
Technical sciences(engineering)	41,495	25.5	61,107	27.3	129,843	36.7	298,811	45.0	409,470	44.1	548,000	46.8
Social sciences	17,030	10.5	24,713	11.0	34,500	9.7	62,318	9.4	102,732	11.1	127,700	10.9
Economics	4,584	2.8	8,247	3.7	13,884	3.9	30,706	4.6	57,518	6.2	80,100	6.8
Geography	2,574	1.6	3,381	1.5	4,274	1.2	5,875	0.9	7,242	0.8	8,300	0.7
Law	1,046	0.6	1,607	0.7	2,249	0.6	3,272	0.5	4,765	0.5	6,300	0.5
Education	8,826	5.4	11,478	5.1	14,093	4.0	22,465	3.4	31,283	3.4	30,200	2.6
Psychology[1]	--	--	--	--	--	--	--	--	1,924	0.2	2,800	0.2
Humanities	28,677	17.6	37,048	16.5	46,679	13.2	73,509	11.1	98,080	10.6	109,900	9.4
History	8,450	5.2	13,444	6.0	16,456	4.6	20,618	3.1	25,138	2.7	28,700	2.5
Philosophy	2,686	1.7	1,861	0.8	3,375	1.0	7,420	1.1	12,039	1.3	15,100	1.3
Philology	13,601	8.4	17,743	7.9	21,234	6.0	37,175	5.6	48,721	5.3	51,600	4.4
Art	3,940	2.4	4,000	1.8	5,614	1.6	8,296	1.2	12,182	1.3	14,500	1.2
Architecture[2]	782	0.5	876	0.4	1,438	0.4	2,003	0.3	2,590	0.3	3,300	0.3
Other(military & military-related)[3]	3,566	2.2	6,514	2.9	7,373	2.1	19,800	3.0	30,663	3.3	39,500	3.4

[1] Psychology was not reported as a separate branch of science until the end of 1968.

[2] Includes history and theory, building design, city planning and landscape architecture, but excludes agricultural engineering, which is included under the technical sciences.

[3] Calculated as a residual except for 1974 for which the figure is reported. See note p. 105.

Note: Data are actual numbers, except for 1974 which are rounded to the nearest hundred.

Source: Data for 1950, 1955, 1960, 1965, and 1970 from Gvishiari, et. al., The Scientific Intelligentsia in the U.S.S.R., Moscow, Progress Publishers, p. 130 (1976); data for 1974 from The National Economy of the U.S.S.R. in 1974, p. 144.

TABLE VI-6

AVERAGE ANNUAL RATE OF GROWTH IN U.S.S.R. SCIENTIFIC WORKERS,
BY BRANCH OF SCIENCE:
1951 to 1974
(In Percent)

	1951 to 1955	1956 to 1960	1961 to 1965	1966 to 1970	1971 to 1974	1951 to 1974
Total	6.6	9.6	13.4	6.9	6.0	8.6
Physical and Life Sciences	5.7	7.5	9.2	6.4	4.7	6.8
Physics and Mathematics	14.5	7.6	17.1	8.3	5.2	10.7
Chemistry	4.9	9.8	5.0	6.4	4.0	6.1
Biology	5.0	6.5	12.4	6.7	5.1	7.2
Geology/Mineralogy	9.3	13.5	9.0	4.3	4.8	8.3
Medicine	3.3	4.8	2.6	6.4	4.3	4.3
Pharmacy	3.6	8.5	3.1	5.9	3.8	5.1
Agriculture	1.4	7.0	8.5	2.8	4.0	4.8
Veterinary Sciences	1.4	6.6	1.7	4.2	4.9	3.7
Technical Sciences (Engineering)	8.0	16.3	18.1	6.5	7.6	11.4
Social Sciences	7.7	6.9	12.5	10.5	5.6	8.7
Economics	12.5	11.0	17.2	13.4	8.6	12.7
Geography	5.6	4.8	6.6	4.3	3.5	5.0
Law	9.0	7.0	7.8	7.8	7.2	7.8
Education	5.4	4.2	9.8	6.9	-0.9	5.3
Psychology[1]	--	--	--	--	9.8	--
Humanities	5.3	4.7	9.5	5.9	2.9	5.8
History	9.7	4.1	4.6	4.0	3.4	5.2
Philosophy	-7.1	12.6	17.1	10.2	5.8	7.5
Philology	5.5	3.7	11.9	5.6	1.4	5.7
Art	0.3	7.0	8.1	8.0	4.4	5.6
Architecture	2.3	10.4	6.9	5.3	6.2	6.2
Other (Military & Military-Related)[2]	12.8	2.5	22.0	9.1	6.5	10.5

[1] Psychology was not reported as a separate branch of science until the end of 1968.

[2] Believed to be primarily military and military-related, as previously indicated.

Source: Calculated on the basis of data contained in Table VI-5.

Power engineering
Metallurgy
Mechanical engineering and instrument construction
Electronics, electrical instrument construction, and automation
Radio engineering and communications
Chemical technology
Timber engineering and the technology of woodpulp, cellulose, and
    paper
Technology of food products
Technology of consumer goods industry
Building and construction
Geodesy and cartography
Hydrology and meteorology
Transportation

In addition, some of the specialties in agriculture are classified as engineering.

While data are reported annually on the distribution of enrollment and graduation in engineering specialties in higher educational institutions, the distribution of "diplomaed engineers" employed in the national economy is not reported by specialty branch. The number of "diplomaed engineers" from 1950 to 1975 is shown in Table VI-7. In 1975, there were more than three and one-half million engineers employed in the USSR, representing almost a ten-fold increase over the number in 1950. The average annual rate of growth throughout the period was 9.3 percent, with the highest rate of increase occurring between 1955 and 1960.

C. The Scientist and Engineer Population in the United States

In the United States, published data on the number of scientists and engineers rely on various definitions, depending on the organization that is collecting the statistics and the nature of the statistical study. One important source, and the study that was principally utilized in the U.S. survey report on "Training and Utilization of Scientific and Engineering Technical Personnel" exchanged with the Soviet Union as part of the cooperative efforts of the Science Policy Working Group, is the National Science Foundation study entitled "The 1972 Scientist and Engineer Population Redefined,"[6] also known as the 1972 Postcensal Survey. The NSF study was based on responses to questionnaires

---

[6]NSF 75-313 and NSF 75-327.

TABLE VI-7

U.S.S.R. EMPLOYED "DIPLOMAED ENGINEERS":
1950 to 1977

	Number in Thousands
1950	400.2
1955	597.8
1960	1,135.1
1965	1,630.8
1970	2,486.5
1974	3.370.0
1975	3,683.0
1977	4,193.0

Sources: 1950, 1960, 1970, and 1975: S. R. Mikulinsky, et. al,, "The
          Training and Utilization of Scientific and Engineering-Technical
          Personnel in the USSR: Part II" Moscow, p. 30 (December 1976);
          1955 and 1965: Trud v SSSR, Statisticheskiy Sbornik, Moscow
          Statistika. 1968, p. 252; National Economy of the USSR in 1974,
          p. 135; National Economy of the USSR in 1977, p. 393.

sent to a sample of persons who, on the basis of self-identification, had been classified by the 1970 Census of Population either as currently employed in scientific, engineering, technical, or related occupations, or as having completed four or more years of higher education, regardless of current occupation. The Postcensal Survey essentially attempted to filter out technicians from the scientist and engineer population as identified by the census, and at the same time to classify as scientists or engineers various individuals who had been trained or otherwise qualified in engineering and science fields but, because they were neither currently employed as scientists or engineers nor unemployed but having last worked in those fields, had been excluded from the scientist and engineer population as identified by the census.

In contrast to the census, which classified scientists and engineers solely on the base of self-identification of current or most recent occupation, the Postcensal Survey relied on three criteria for redefining the scientist and engineer population: professional self-identification, present and past employment, and education (in terms of level of highest degree held or years of education obtained and major field of study.)[7] Thus, for example, an individual who identified himself as an engineer and had a bachelor's degree or higher in a field of engineering was classified as an engineer regardless of whether or not he was currently employed as an engineer, as was an individual who identified himself as an engineer, had from two to four years of college but no degree, but had at least six years of prior job experience as an engineer. On the other hand, no individual who was currently employed in a field other than engineering but who also failed to identify himself as an engineer was classified as an engineer, regardless of his educational background. Overall, over 90 percent of the total scientist and engineer population as redefined in the Postcensal Survey had a bachelor's or higher degree, and there was a high level of correlation between the field of science or engineering in which the individual was classified and the major field of study in which his highest degree was held. The criteria provided for mutually exclusive populations so that an individual could not be classified in more than one field of science or engineering.

The Postcensal Survey provides especially good insights into career patterns and mobility of scientific and technical personnel, the extent to which individuals trained in scientific, engineering, and technical fields are working in occupations apparently unrelated to their training,

---

[7]*The 1972 Scientist and Engineer Population Redefined*, Volume 2, National Science Foundation, NSF 75-327 (September 1975).

and the activities and duties that are actually performed by persons classified in scientific and technical occupations. For this reason, these data are used as the basis for the discussion of the mobility of scientists and engineers in the United States presented in Chapter VIII, as well as for some of the discussion of technicians presented in Chapter III. However, the wide range of possible combinations which satisfy the various criteria for inclusion among the scientist and engineer population in the Postcensal Survey makes this a particularly difficult study to utilize in comparison with other statistical reports which rely on alternative definitions, as do the Soviet statistics.

Data published by the Bureau of Labor Statistics (BLS), on the other hand, are based on a more easily definable criterion for inclusion of an individual in the scientist and engineer population, and thus are more suitable for comparison with statistics based on alternative definitions. BLS statistics classify individuals as scientists or engineers only if they are actually engaged in scientific or engineering work at a level which requires a knowledge of the science or engineering field in which they are engaged equivalent to that acquired through completion of a four-year college course with a major in that field of science or engineering, regardless of whether or not they hold a college degree. Persons trained in the sciences or engineering but currently employed in positions not requiring the use of such training are excluded.[8] Because of its more easily definable criterion for the scientist and engineer population, the BLS statistics will be used throughout the remainder of this section and in the following section comparing the population of scientists and engineers in the United States with that in the Soviet Union.

Data on the number of engineers, physical scientists, mathematicians, and life scientists in the United States from the year 1950 to 1970 and for 1974 are shown in Table VI-8. Table VI-9 shows their employment by sector and individual industry for the same years. During the 1950s, the number of scientists and engineers in the United States doubled, rising from about 600,000 to nearly 1.2 million. In the 1960s, employment increased by almost as much in absolute terms, but the relative gain was only half of that experienced between 1950 and 1960. Furthermore, in the 1960s the number of scientists grew significantly faster than the number of engineers (75 and 38 percent, respectively), partially as a result of substantial gains in social science

[8]"Detailed Reporting Instructions," *A Survey of Scientific and Technical Personnel in Industry, in 1969*, U.S. Department of Labor, Bureau of Labor Statistics, BLS 2716A (1971).

TABLE VI-8

U.S. SCIENTISTS AND ENGINEERS, BY FIELD: 1950-70 and 1974

(In Thousands)

Year	Total Scientists & Engineers	Engineers	Physical scientists Total	Chemists	Physicists	Geologists	Other	Mathematicians	Life scientists Total	Agricultural	Biological	Medical
1950	556.7	408.0	89.1	51.9	14.0	13.0	10.2	13.8	45.6	16.9	19.9	8.8
1951	611.8	450.6	97.6	56.8	15.2	13.3	12.3	14.7	48.9	18.2	21.2	9.5
1952	685.9	507.5	108.6	62.9	16.7	13.8	15.2	16.1	53.7	20.4	23.0	10.3
1953	748.7	556.2	118.3	67.9	18.0	15.5	16.9	17.7	56.6	21.5	24.1	11.0
1954	783.7	581.2	124.2	71.6	19.1	16.1	17.4	19.5	58.9	21.7	25.5	11.7
1955	812.6	601.4	128.3	73.9	19.9	17.1	17.4	21.1	61.8	22.2	27.3	12.3
1956	873.7	646.4	137.0	79.2	21.4	17.9	18.5	23.1	67.3	23.7	29.9	13.7
1957	958.9	707.9	148.4	84.5	23.7	19.6	20.6	26.1	76.7	25.6	34.8	16.3
1958	1001.2	730.3	157.4	90.6	26.1	20.1	20.6	28.5	84.9	27.3	39.0	18.6
1959	1057.9	768.0	166.2	95.4	28.6	20.9	21.3	31.7	92.0	29.5	42.5	20.0
1960	1104.0	801.1	172.0	99.7	29.8	20.4	22.1	34.2	96.7	30.4	44.8	21.5
1961	1151.5	833.3	178.8	102.8	31.6	20.6	23.8	36.3	103.3	32.3	46.9	24.1
1962	1210.3	873.2	186.0	106.8	33.9	21.1	24.2	39.8	111.5	35.3	49.0	27.2
1963	1280.8	922.7	194.1	110.0	36.3	22.5	25.3	43.6	120.3	38.5	51.3	30.5
1964	1327.0	945.5	203.7	115.0	39.0	23.4	26.3	47.2	130.5	41.5	54.4	34.6
1965	1366.3	969.8	209.2	116.7	39.9	25.5	27.1	50.3	136.9	44.1	55.6	37.2
1966	1417.5	999.6	217.0	119.6	42.1	26.2	29.1	53.9	147.0	46.9	56.9	43.2
1967	1476.7	1037.7	225.8	122.8	44.4	28.4	30.2	61.9	151.1	46.5	62.6	42.0
1968	1525.0	1062.4	236.1	127.3	46.2	29.0	33.6	67.1	159.4	47.2	65.8	46.4
1969	1567.7	1085.0	243.8	131.0	48.4	29.4	35.0	73.0	165.9	47.5	67.7	50.7
1970	1594.7	1098.2	248.8	132.9	49.1	30.6	36.2	74.3	173.4	49.3	71.1	53.0
1974	1631.9	1114.0	251.6	134.5	47.5	31.5	38.1	77.5	188.8	47.7	76.8	64.3

Note: Detail may not add to totals due to rounding; 0.0 is less than 50.

Source: (a) Data for 1974 in Occupational Outlook Handbook, 1976-77 Edition, Bureau of Labor Statistics Bulletin 1875.

(b) Data for 1950-70 in Employment of Scientists and Engineers, 1950-70, Bureau of Labor Statistics Bulletin 1781.

TABLE VI-9

U.S. SCIENTISTS AND ENGINEERS, BY SECTOR AND INDIVIDUAL INDUSTRY: 1950-70 AND 1974
(In Thousands)

Page 1 of 2

SECTOR	1950	1951	1952	1953	1954	1955	1956	1957	1958	1959	1960	1961
All sectors	556.7	611.8	685.9	748.7	783.7	812.6	873.7	958.9	1001.2	1057.9	1104.0	1151.5
Private industry	396.3	439.4	500.1	557.5	590.0	610.1	656.3	719.6	743.5	778.3	812.0	842.2
Manufacturing	245.2	275.9	326.8	371.0	393.2	399.6	427.3	478.1	496.5	521.4	550.1	577.1
Ordnance	2.1	3.0	10.4	16.2	16.1	14.8	14.6	18.1	17.8	29.2	34.1	40.6
Food	9.7	10.6	11.2	11.6	12.1	12.4	12.8	13.1	13.6	14.2	14.8	15.4
Textiles and apparel	3.1	3.4	3.5	3.6	3.6	3.8	3.8	3.9	3.9	4.0	4.2	4.0
Lumber and furniture	2.2	2.5	2.6	2.7	2.7	2.7	2.9	3.1	3.1	3.2	3.3	3.4
Paper	5.6	6.4	6.9	7.4	7.7	8.3	9.0	9.6	10.0	10.3	10.9	11.8
Chemicals	39.4	45.3	50.4	56.1	60.2	60.4	63.3	68.6	73.9	74.5	77.3	79.4
Petroleum refining	8.7	9.5	10.7	11.7	12.5	13.0	13.2	13.3	14.0	14.0	14.6	14.2
Rubber	6.1	6.6	7.3	7.9	8.0	8.1	8.5	9.0	9.1	9.4	9.7	10.2
Stone, clay, and glass	5.4	6.0	6.4	6.9	7.1	7.5	8.2	8.6	9.1	9.5	10.0	10.2
Primary metals	15.4	17.6	19.2	20.7	21.3	20.4	22.3	24.9	25.8	26.7	28.2	30.3
Fabricated metals	15.5	17.7	18.9	20.2	20.4	21.0	22.7	24.0	24.1	24.3	25.2	25.9
Machinery	33.6	38.0	43.0	45.5	47.4	47.5	49.9	55.9	56.5	59.1	62.6	67.2
Electrical equipment	44.9	48.5	57.3	66.5	72.6	74.9	82.1	97.0	107.4	108.7	118.9	128.5
Motor vehicles	17.6	18.5	18.8	19.2	19.7	19.8	20.6	20.5	21.0	21.4	22.2	22.5
Aircraft	17.6	22.4	37.7	50.0	55.7	58.8	65.3	77.7	74.6	78.0	76.3	74.1
Other transportation equipment	2.8	2.8	3.2	3.7	3.6	3.3	3.7	4.3	4.7	4.7	4.7	4.8
Professional and scientific instruments	10.5	12.0	14.0	15.7	17.0	17.3	18.6	20.5	21.8	24.0	26.6	28.5
Miscellaneous manufacturing	4.9	5.1	5.3	5.4	5.5	5.6	5.8	6.0	6.1	6.2	6.3	6.1
Nonmanufacturing	151.1	163.5	173.3	186.5	196.8	210.5	229.0	241.5	247.0	256.9	261.9	265.1
Petroleum extraction	14.5	14.5	15.1	17.5	18.7	20.0	21.5	24.0	24.5	24.4	23.3	23.3
Mining	5.8	5.8	6.4	6.1	5.9	5.9	6.1	6.4	6.8	6.9	6.7	6.8
Construction	24.6	29.3	31.4	33.5	33.8	37.3	39.5	41.0	42.1	43.6	45.0	45.5
Railroads	5.3	5.4	5.4	5.4	5.3	5.3	5.3	5.2	5.1	5.0	5.0	4.9
Other transportation	3.5	3.6	3.7	3.8	3.9	4.0	4.1	4.2	4.3	4.6	4.7	5.0
Telecommunications	6.3	6.6	7.2	7.7	8.0	8.2	8.6	8.9	8.9	9.1	9.2	9.4
Radio and TV	1.5	1.8	1.9	1.9	2.5	3.8	4.0	4.1	4.3	4.4	4.4	4.5
Public utilities	10.7	11.8	11.8	13.5	15.1	16.8	17.3	19.0	20.3	20.8	21.6	22.9
Miscellaneous business services	23.1	24.4	25.8	28.5	29.8	29.8	32.3	34.6	37.3	41.9	44.3	45.7
Medical and dental laboratories	0.5	0.6	0.6	0.7	0.8	0.8	0.9	0.9	0.9	1.0	1.0	1.0
Engineering and architectural services	41.3	45.4	49.3	53.4	57.5	61.3	71.0	74.6	73.2	74.8	75.6	74.4
Other nonmanufacturing	14.0	14.3	15.0	15.6	16.6	17.3	18.4	18.6	19.3	20.4	21.0	21.7
Government	106.5	119.5	132.1	135.3	132.9	135.7	142.4	151.3	155.4	165.0	168.5	176.8
Federal	59.7	70.7	82.2	84.9	79.7	81.5	85.4	90.1	91.7	97.7	97.9	102.2
State	26.4	27.8	29.5	29.4	31.5	31.5	33.4	35.3	36.7	39.3	42.0	44.8
Local	20.4	21.0	20.4	21.7	21.7	22.7	23.6	25.9	27.0	28.0	28.6	29.8
Colleges and universities	50.3	49.2	49.4	51.2	55.8	61.2	69.4	81.5	95.3	106.7	114.8	122.5
Nonprofit institutions	3.6	3.7	4.3	4.7	5.0	5.6	5.6	6.5	7.0	7.9	8.7	10.0

Note: Detail may not add to totals due to rounding; 0.0 is less than 50.

TABLE VI-9 (continued)

U.S. SCIENTISTS AND ENGINEERS, BY SECTOR AND INDIVIDUAL INDUSTRY: 1950-70 AND 1974

(In Thousands)

Page 2 of 2

SECTOR	1962	1963	1964	1965	1966	1967	1968	1969	1970	1974
All sectors	1210.3	1280.8	1327.0	1366.3	1417.5	1476.7	1525.0	1567.7	1594.7	1631.9
Private industry	878.3	925.2	946.6	968.0	1004.2	1044.2	1069.8	1100.2	1111.2	1131.7
Manufacturing	603.0	638.1	643.7	658.0	681.5	714.5	729.6	734.4	733.2	724.2
Ordnance	47.0	52.8	54.6	58.5	59.9	60.9	62.4	63.2	52.7	43.8
Food	14.9	15.3	14.1	14.1	14.0	14.0	14.4	14.7	15.2	15.2
Textiles and apparel	4.1	4.3	5.1	5.8	6.3	5.6	5.8	5.8	5.9	6.0
Lumber and furniture	3.4	3.4	3.5	3.7	4.0	4.3	4.8	5.0	5.0	5.8
Paper	12.6	12.6	12.8	13.0	13.3	14.0	14.6	14.7	15.0	15.2
Chemicals	83.1	85.0	90.8	91.8	97.7	99.6	100.2	103.5	107.3	110.1
Petroleum refining	14.4	14.6	14.8	14.0	13.8	14.7	13.2	13.2	13.2	13.2
Rubber	10.3	11.1	11.5	12.2	13.5	14.1	14.5	14.2	15.0	16.1
Stone, clay, and glass	10.5	10.7	10.5	10.5	11.2	12.0	11.7	12.1	12.3	12.4
Primary metals	30.3	28.7	27.0	27.1	27.7	29.2	28.5	30.9	29.8	30.0
Fabricated metals	26.0	27.0	27.3	28.9	30.2	31.8	29.0	31.2	30.2	31.9
Machinery	70.4	75.7	79.9	80.0	81.6	88.6	91.3	89.8	94.1	98.6
Electrical equipment	133.8	146.1	141.0	142.2	146.2	152.5	160.4	158.8	161.8	162.5
Motor vehicles	22.8	24.6	27.3	29.6	32.1	32.4	31.8	32.4	33.9	38.4
Aircraft	77.9	83.0	80.9	82.3	83.3	92.8	99.9	98.2	91.8	71.5
Other transportation equipment	4.8	5.0	4.9	5.2	5.5	5.9	5.8	5.4	6.1	6.7
Professional and scientific instruments	30.4	31.5	31.2	32.8	34.8	35.8	35.4	35.7	37.5	40.3
Miscellaneous manufacturing	6.3	6.7	6.5	6.5	6.4	6.3	5.9	6.2	6.4	6.5
Nonmanufacturing	275.3	287.1	302.9	309.7	322.7	329.7	340.2	365.8	378.0	407.5
Petroleum extraction	22.9	23.8	23.8	24.2	23.9	24.6	25.0	25.8	24.7	24.9
Mining	6.4	6.8	6.5	6.9	7.2	7.1	7.1	7.6	7.8	8.8
Construction	46.9	48.2	47.4	47.1	51.6	48.1	47.2	53.6	55.2	54.9
Railroads	4.8	4.5	4.9	4.4	4.3	4.5	4.5	4.4	4.4	4.4
Other transportation	5.0	4.8	5.0	5.4	5.4	5.3	5.4	5.6	5.8	6.4
Telecommunications	9.7	9.8	10.5	11.8	12.1	12.9	15.0	15.7	16.1	18.2
Radio and TV	4.5	4.5	4.7	5.1	5.1	5.3	5.6	5.8	6.0	6.0
Public utilities	24.0	24.6	25.3	26.7	27.1	27.0	27.5	28.3	30.1	35.0
Miscellaneous business services	47.7	53.1	56.0	55.1	57.0	61.1	65.9	74.5	76.1	76.0
Medical and dental laboratories	1.1	1.1	1.2	1.3	1.4	1.5	1.5	1.4	1.5	1.7
Engineering and architectural services	77.3	79.7	87.9	89.2	93.0	96.2	97.4	101.9	107.6	123.0
Other nonmanufacturing	25.0	26.2	29.7	32.5	34.7	36.0	38.2	41.1	42.7	48.2
Government	188.6	199.3	207.0	214.8	218.5	221.0	231.9	233.0	238.9	237.5
Federal	110.6	120.0	126.4	132.1	134.2	136.0	145.4	146.3	150.4	144.4
State	47.4	47.9	48.5	49.7	50.4	51.2	52.8	53.2	54.3	57.5
Local	30.6	31.4	32.1	33.0	33.7	33.8	33.7	33.5	34.2	35.6
Colleges and universities	130.9	141.6	158.2	167.8	178.1	194.3	205.8	217.1	228.2	247.2
Nonprofit institutions	12.5	14.7	15.2	15.7	16.7	17.2	17.5	17.4	16.4	15.5

Note: Detail may not add to totals due to rounding; 0.0 is less than 50.
Source: (a) Data for 1974 in Occupational Outlook Handbook, 1976-77 Edition, Bureau of Labor Statistics Bulletin 1875; (b) Data for 1950-70 in Employment of Scientists and Engineers, 1950-70, Bureau of Labor Statistics Bulletin 1781.

fields. Beginning in 1969, growth in employment of scientists and engineers slowed and then remained relatively level until 1974.

D. U.S.-Soviet Comparison

Soviet data on both "scientific workers" and "diplomaed engineers" include fields or occupations not classified as science or engineering in the United States. Many fields in the humanities, for example, are included among the science fields, while engineering includes fields such as geodesy and cartography, hydrology and meteorology, and some specialties in agriculture that are not considered as engineering in the United States. As data are available for some years showing the breakdown of the number of scientific workers by field,[9] employment in those fields that are not considered as science or engineering in the United States can be subtracted from the total number of scientific workers for comparison purposes. Soviet data on diplomaed engineers employed in the economy, on the other hand, are not broken down by field of specialization and can thus be used only in the aggregate when compared with U.S. data. The extent to which Soviet aggregate data on diplomaed engineers are inflated relative to U.S. data on engineers employed in the economy can be inferred, however, from Soviet data on annual graduation of engineers from higher educational institutions, for which breakdowns by field are available. In 1974, about 20 percent of all Soviet engineering graduates were trained in fields not classified as engineering in the United States.

Another problem with the use of Soviet data on diplomaed engineers for comparison with U.S. statistics on employment of engineers is that noted above: the Soviet data include all persons who have received a diploma in engineering fields, regardless of whether they are actually working as engineers. The great number of Soviet engineering graduates who received their training in evening and correspondence programs, the quality of which is generally conceded to be well below that of formal engineering degree programs in the United States,[10] presents an additional problem with using Soviet data on engineers in a comparison with U.S. statistics. It should also be noted, however, that the bachelor of science in engineering degree in the United States covers a range of engineering qualifications and there is substantial variability in quality of programs and ability of students; new engi-

[9] 1974 is the latest year for which such a breakdown is available.
[10] See Chapter IV.

neering graduates are therefore employed in a range of positions involving various levels of technical expertise and responsibilities.

While the differences in definitions used for statistical reporting and qualitative differences in engineering degrees such as those noted above require that direct comparisons of the number of scientists and engineers in the United States and the Soviet Union be viewed with caution, the data nevertheless provide the basis for some assessment of relative changes in emphasis on science and engineering manpower in the two countries. Table VI-10 shows the number of natural scientists and engineers in the United States and the Soviet Union from 1950 to 1974. In 1950, the total number of natural scientists and engineers was about 20 percent greater in the United States than in the USSR; by 1974, however, the USSR had over twice as many natural scientists and engineers as the United States. This change in the relative number of scientists and engineers in the United States and the Soviet Union was largely the result of the enormous increase in the number of engineers in the USSR during this period. While in 1950, there were approximately the same number of engineers in the two countries, by 1974 the USSR had over three times as many engineers as did the United States.

In the natural sciences, while there were more than twice as many scientists in the United States as in the Soviet Union in 1950, by 1974 this discrepancy had become somewhat less pronounced so that there were only about one and one-half times as many natural scientists in the United States as in the Soviet Union. The fields of natural science in which the Soviet Union showed the greatest increase relative to the United States were geology, in which the average annual rate of growth was 8.3 percent in the Soviet Union compared to 3.8 percent in the United States, and physics and mathematics, where the average annual rate of growth was 10.7 percent in the Soviet Union compared to 6.5 percent in the United States. In the medical sciences, on the other hand, the United States during this period showed a much higher average annual rate of growth (8.6 percent) than did the Soviet Union (4.3 percent); thus, while the Soviet Union had almost two-and-one-half times as many medical scientists as did the United States in 1950, by 1974 the United States had slightly more medical scientists than did the Soviet Union.

Figure VI-1 shows the distribution of natural scientists among fields of science for the United States and the Soviet Union in 1950 and 1974. The difference in the number of natural scientists in the two countries was most pronounced in the field of chemistry in both 1950

TABLE VI-10

US AND USSR NATURAL SCIENTISTS AND ENGINEERS:
1950 to 1974
(In Thousands)

	1950		1955		1960		1965		1970		1974		Average Annual Rate of Growth 1951 to 1974	
	US	USSR	US	USSR	US	USSR	US	USSR	US	USSR	US	USSR	US	USSR
Total natural scientists and engineers	556.7	468.5	812.6	688.6	1104.0	1265.4	1366.3	1834.5	1594.7	2765.2	1631.9	3704.7	4.6	9.0
Natural scientists	148.7	68.3	211.2	90.8	302.9	130.3	396.5	203.7	496.5	278.7	517.9	334.7	5.3	6.8
Physicists and mathematicians	27.8	10.2	41.0	20.1	64.0	29.0	90.2	63.9	123.4	95.3	125.0	116.9	6.5	10.7
Chemists	51.9	12.9	73.9	16.4	99.7	26.2	116.7	33.5	132.9	45.8	134.5	53.7	4.0	6.1
Biologists	19.9	8.6	27.3	11.0	44.8	15.1	55.6	27.1	71.1	37.3	76.8	45.5	5.8	7.2
Geologists	13.0	3.6	17.1	5.7	20.4	10.7	25.5	16.4	30.6	20.3	31.5	24.5	3.8	8.3
Other physical scientists[1]	10.2	--	17.4	--	22.1	--	27.1	--	36.2	--	38.1	--	5.6	--
Agricultural scientists	16.9	11.9	22.2	12.8	30.4	18.0	44.1	27.1	49.3	31.1	47.7	36.5	4.4	4.8
Medical scientists	8.8	21.0	12.3	24.8	21.5	31.4	37.2	35.8	53.0	48.8	64.3	57.6	8.6	4.3
Engineers	408.0	400.2	601.4	597.8	801.1	1135.1	969.8	1630.8	1098.2	2486.5	1114.0	3370.0	4.3	9.3

1 "Including metallurgists and other specialities classified by the USSR as engineering," Bronson, op. cit., p. 589 fn.

Source: Data from Tables VI-5, VI-7, and VI-8.

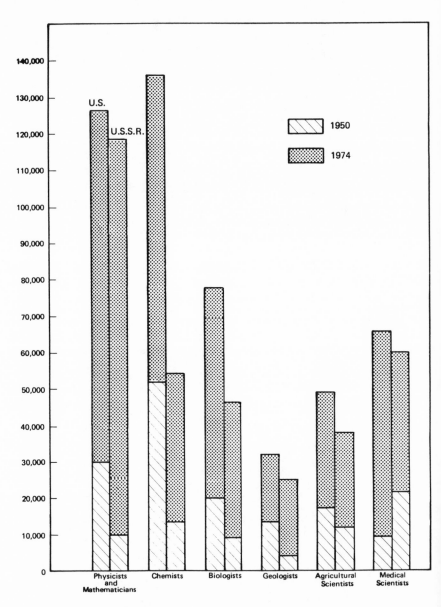

Source: See Table VI-1

Figure VI-1   U.S. AND U.S.S.R. NATURAL SCIENTISTS: 1950 AND 1974

and 1974, while it was least pronounced in agriculture in 1950 and in the medical sciences, geology, and physics and mathematics in 1974.

Table VI-11 shows the average annual rate of growth of science and engineering manpower in the United States and the Soviet Union from 1950 to 1974. Between 1950 and 1974, the average annual rate of growth in the number of natural scientists was slightly higher in the Soviet Union (6.8 percent) than in the United States (5.3 percent). The average annual rate of growth in the number of engineers, however, was more than twice as high in the Soviet Union (9.3 percent) as in the United States (4.3 percent). In the United States, the rate of growth of natural scientists exceeded the rate of growth of engineers, while in the Soviet Union, the reverse occurred.

As the above figures indicate, the growth in scientific and engineering personnel in the Soviet Union has been quite dramatic. However, there is significant evidence from both Soviet and Western sources of misutilization or underemployment of engineers, scientists, and other skilled R&D personnel in many fields of training. Problems have been noted in the utilization of engineers to perform tasks that could be accomplished by lesser qualified and auxiliary personnel. Engineers in Leningrad industrial enterprises were reported to waste 22.2 percent of their time in the performance of such functions.[11] Another criticism leveled at the utilization of professional manpower in the USSR is that too many engineers and scientists are employed in administrative capacities because of a shortage of qualified economists and managerial personnel in industry. In addition, many diplomaed engineers are employed in fields other than that in which they were trained, due to lack of vacancies in their fields of specialization. Prior job experience requirements for certain positions have contributed to the occupation of positions requiring higher education by experienced individuals without formal training.

One of the reasons cited for misassignments such as those noted above is the tendency of industrial plants and enterprises to inflate their statements of requirements for specialists in order to avoid possible shortages in the future. It has been reported that the "hidden reserves" resulting from this tendency on the part of industrial enterprises comprises 10 to 15 percent of all workers in industry.[12]

---

[11]A. Grigor'yev, "Whom Shall We Train in the Higher Educational Institution?" *Leningradskaya Pravda*, pp. 2–3 (April 15, 1977).

[12]Ye. Kasimovskiy, "Prospects for the Formation and Utilization of Labor Resources in the USSR," *Ekonomicheskiye Nauki*, No. 8, pp. 33–45 (August 1973), cited by Jill E.Heuer, "Soviet Professional, Scientific, and Technical Manower," Defense Intelligence Agency, DST-1830S-049-79, p. 5 (May 1979).

TABLE VI-11

U.S. AND U.S.S.R AVERAGE ANNUAL RATE OF GROWTH
OF SCIENCE AND ENGINEERING MANPOWER: 1951 TO 1974
(In Percent)

	Natural Scientists		Engineers	
	U.S.	U.S.S.R	U.S.	U.S.S.R.
1951 to 1955	7.8	5.9	8.1	8.4
1956 to 1960	7.5	7.5	5.9	13.7
1961 to 1965	5.5	9.3	3.9	7.5
1966 to 1970	4.6	6.5	2.5	8.8
1971 to 1974	1.1	4.7	0.4	7.9
1951 to 1974	5.3	6.8	4.3	9.3

Source: Calculated on the basis of data from Table VI-10.

Another factor that has been noted as contributing to the misutilization of Soviet professional manpower is the ratio of professionals to technicians and other semiprofessional personnel. As previously noted, a ratio of 3 to 4 technicians per engineer is considered desirable for industry, while the actual ratio prevailing in 1977 was 1.6:1. With a lack of adequate support personnel, highly trained professionals must devote a significant portion of their time to routine functions and administrative duties. Other inflexibilities affecting the effective utilization of scientific and engineering personnel in the Soviet Union have been attributed to restrictions on mobility, over-rigid adherence to quotas, the lack of incentives, and the traditional separation of R&D and production under different management organizations.[13]

A different view of the need for such large numbers of scientists and engineers in the Soviet economy is taken by V.G. Afanasyev, a leading Party theoretician and proponent of the "scientific management of society":

> The general economic effect of education is exceptionally high. Education is now considered one of the most effective forms of investment. . . . Take, for example, the question of the number of specialists with higher education, particularly engineers. It is said that the USSR has a large surplus of engineers. If we consider only the demands of the present, the point has a certain validity. Quite often one finds a fully trained engineer doing work that does not require such a high qualification. On the other hand, if we take the long view and consider the needs of the future, the idea that we have a surplus of engineers falls to the ground.
>
> Modern production is now in the process of gradual transition from mechanization to automation, which is bound to boost the importance of skilled workers and specialists . . .[14]

The clearly stated approach of the Soviet leadership to respond to increasing limitations on natural resources and decrease in population growth relies on qualitative changes in production—changes largely brought about through the development and utilization of scientific and technological skills and abilities of the labor force—to effect quantitative improvement in the economic sphere.[15] Seen in the light of this present broad strategy, the Soviet commitment to the development of

[13]"Engineering Manpower in the Soviet Union," *Engineering Manpower Bulletin*, Engineers Joint Council, p. 1 (1977).

[14]V.G. Afanasyev, *The Scientific and Technological Revolution—Its Impact on Management and Education*, Moscow, Progress Publishers, p. 293–95 (1975).

[15]See Catherine P. Ailes, James E. Cole, and Charles H. Movit, "Soviet Economic Problems and Technological Opportunities," *Comparative Strategy*, Vol. 1, No. 4 (1979).

cadres of scientists and engineers despite their underemployment in the current economic structure is less surprising. As Afanasyev points out, the technological transformation of the economy would eliminate the surplus of such technically trained personnel. There are at the present time, however, no clear signs of any major reorganization of the Soviet economy such as would probably be required to bring about such a technological transformation.

CHAPTER VII

Utilization of Scientists and Engineers in R&D Employment Sectors

A significant amount of attention in the United States has been devoted to the definition of R&D activity and the internal divisions differentiating basic from applied research and research as a whole from development. The U.S. National Science Foundation (NSF) publishes a variety of annual publications and special reports providing quantitative data about the structure and growth of employment in R&D as well as R&D expenditures. By contrast, the Soviet system has been less concerned with clarifying the borderline between R&D and other activities or between the stages of the R&D cycle. The concept used in Soviet statistical reporting that is generally regarded by Western analysts as a rough approximation of the concept of "scientists and engineers engaged in R&D" as used by NSF is that of "scientific workers." The "scientific worker" concept, however, differs from the U.S. concept of scientists and engineers engaged in R&D in that it is wider in certain respects and narrower in others.[1] Efforts have been made by a number of Western analysts, however, to adjust the data provided in the Soviet national statistics to arrive at comparability in scope with U.S. R&D employment aggregates.

Apart from differences in the way R&D activity is conceived and defined in the United States and the USSR, there are also major differences in the types of organizations in the two countries that are engaged in this type of activity. In both the United States and the Soviet Union, the research and development enterprise is scattered throughout a number of different kinds of institutions, which differ in their goals, mechanisms of financing, and major functions. In the

[1]See Chapter VI, Section B, for a discussion of the Soviet "scientific worker" series.

Soviet Union, organizations that perform research and development fall into the following sectors:

- Academies of sciences institutes
- Scientific organizations of higher educational institutions
- Branch scientific organizations
- Research units of industrial enterprises.

In the United States, the following broad sectors are generally used in classifying the diversified organizations that perform R&D:

- Federal government
- Universities and colleges
- Industry
- Nonprofit institutions

While a sectoring of the R&D effort by type of performer is possible in both the United States and the USSR, the individual sectors in the two countries are not directly comparable due to great differences in the institutional characteristics of the two systems. While Soviet branch institutes are somewhat analogous to the U.S. industry sector, there is no analogue in the United States to the research institutes of the Soviet Academy of Sciences. In addition, while the R&D performer sectors of both countries include a higher educational institution category, the research and development efforts of these institutions differ between the two countries in several important respects.

This chapter will first discuss aggregate employment of scientists and engineers in R&D in the two countries, based on the concept of R&D utilized by the U.S. National Science Foundation and various adjustments that have been made of Soviet R&D manpower data to approximate NSF concepts. The chapter will then describe the major R&D performer categories in the Soviet economy and discuss employment in these sectors, using data from the scientific workers series as a rough approximation of the U.S. concept of scientists and engineers engaged in R&D (generally without adjusting the series for closer comparability to the U.S. statistics). Finally, the broad categories of U.S. organizations that perform R&D and the employment of scientists and engineers within these sectors will be described.

A. Scientists and Engineers Engaged in R&D

The National Science Foundation defines research as "systematic, intensive study directed toward fuller scientific knowledge of the subject studied'' and development as ''the systematic use of scientific knowledge directed toward the production of useful materials, devices, systems or methods, including design and development of prototypes and processes.''[2] NSF figures on scientists and engineers engaged in R&D are based on the number of full-time and part-time employees conducting research and development or R&D management activities, reduced to full-time equivalents. The minimum standard for inclusion of scientists or engineers in the series is described as ''the performance of professional scientific or engineering work in research and development, requiring a bachelor's degree, or its equivalent, in science or engineering.'' Fields of science included in the coverage are the natural sciences (life, physical, and engineering) and the social sciences and psychology, except in the case of the industry sector where coverage is limited to the natural sciences.

Table VII-1 shows NSF data on U.S. full-time equivalent scientists and engineers engaged in R&D and various adjustments of Soviet scientific manpower data to approximate NSF concepts. Estimates of Soviet R&D employment presented in Column A were prepared by Jill E. Heuer, updating a series of reports on Soviet professional scientific and technical manpower published by the Defense Intelligence Agency.[3] Estimates in Columns B_1 and B_2 were prepared by Robert W. Campbell for the National Science Foundation.[4] The large difference between the estimates in Column A and those in Columns B_1 and B_2 are due to basic differences in the methodological approaches utilized by Heuer and Campbell.

Heuer's estimates basically represent adjustments of the Soviet scientific workers series to exclude scientific workers in the social sciences and humanities and to reduce scientific teaching personnel and scientific workers in industrial plants, project and project design organizations to a full-time equivalent employment basis. In addition, a full-time equivalent figure for graduate students in the natural sciences and engineering has been added to the adjusted scientific workers series data. This approach of adjusting the Soviet scientific workers series to approximate the NSF concept of scientists and engineers engaged in R&D is similar to the methodology suggested by Louvan E. Nolting

---

[2]See definitions in *National Patterns of R&D Resources*, National Science Foundation, annual series.
[3]Jill E. Heuer, *Soviet Professional, Scientific, and Technical Manpower*, Defense Intelligence Agency, DST-1830S-049-79 (May 1979).
[4]Robert W. Campbell, *Reference Source on Soviet R&D Statistics, 1950-1978*, National Science Foundation (1978).

TABLE VII-1

U.S. AND U.S.S.R. FULL-TIME EQUIVALENT
SCIENTISTS AND ENGINEERS ENGAGED IN R&D: 1960 to 1977
(In Thousands)

	U.S.	U.S.S.R. (est.)		
		A	B_1	B_2
1960	380.9	235	295.7	328.2
1961	425.7	275	340.5	374.8
1965	494.5	435	521.8	561.4
1970	546.5	575	733.3	806.9
1971	526.4	618	804.2	881.8
1972	518.3	647	862.5	950.1
1973	518.4	674	966.7	1072.1
1974	525.1	710	995.8	1108.0
1975	534.9	739	1061.2	1187.6
1976	549.2	755	1113.7	1254.5
1977	573.9	NA	1147.7	1299.1

SOURCES: U.S., 1960:   D. Bronson, "Scientific and Engineering
Manpower in the U.S.S.R. and Employment in R&D,"
Soviet Economic Prospects for the Seventies, Joint
Economic Committee, U.S. Congress, p. 586 (1973);
U.S., all other years: National Patterns of Science
and Technology Resources 1980, NSF 80-308, p. 33 and
National Patterns of R&D Resources 1953-77, NSF 77-310,
p. 34; U.S.S.R. Estimates, Column A: Jill E. Heuer,
Soviet Professional Scientific and Technical Manpower,
DIA, DST-18305-049-79, p. 88 (May 1979); Columns B_1 + B_2
Robert W. Campbell, Reference Source on Soviet R&D
Statistics, 1950-1978, NSF, p. 38 (1978).

and Murray Feshbach in a recent study of Soviet R&D employment prepared for the Joint Economic Committee.[5] Nolting and Feshbach actually apply this methodology to data for only one year (1970), but the estimate they arrive at for full-time equivalent R&D scientists and engineers less employment in the social sciences and humanities is close to the estimate made by Heuer for that year: 590,760 compared to 575,000. When the Soviet data are adjusted to exclude only the humanities (a closer approximation to the NSF definition) rather than both the social sciences and the humanities as Heuer does, the estimate arrived at by Nolting and Feshbach for 1970 is 661,929.

The estimates prepared by Campbell are significantly higher than those based almost exclusively on the scientific workers series. Campbell believes that the scientific workers series is too small because it excludes a number of specialists with higher education who appear to be engaged in experimental design work (OKR in its Soviet acronym), which would often be considered development in U.S. concepts. To account for such specialists, Campbell uses Soviet data on specialists with higher education employed in the ''science'' sector (which excludes research units of higher educational institutions except for those that are organizationally distinct as well as research units subsidiary to industrial enterprises), adjusted to exclude those specialists that are already included among scientific workers. The difference between estimates in Column B^1 and those in Column B^2 is that the former include 40 percent of specialists with higher education in the science sector not already included among scientific workers, while the latter estimates include 60 percent of such specialists. If such specialists were subtracted from Campbell's estimates of Soviet full-time equivalent scientists and engineers engaged in R&D, the resulting number for 1970 is 586,100, a figure within the range of Heuer's estimate of 575,000 and Nolting and Feshbach's estimate of 590,760 for that year.[6]

[5]Nolting and Feshbach, op. cit., p. 753.

[6]It should be noted that there are other respects in which estimates made by Heuer, Campbell, and Nolting and Feshbach differ. Most significant in terms of these differences in methodology are branches of science that are included as science or engineering and the way in which graduate student participation in R&D is accounted for.

The branches of science that are included as natural science and engineering fields in Heuer's estimates are: physics and mathematics, chemistry, biology, geology and mineralogy, agriculture and veterinary sciences, medicine and pharmacy, and technical sciences. Full-time equivalent graduate student participation in R&D is estimated as 50 percent of full-time aspirants.

Campbell's estimates include as science and engineering fields those branches of science included by Heuer, as well as geography, economics, and psychology to approximate NSF's inclusion of social science R&D except for the small portion performed by industry. Full-time equivalent graduate student participation in R&D is estimated as 10 percent of full-time aspirants.

Branches of science included by Nolting and Feshbach in making their estimates which exclude both social sciences and humanities are the same as those included by Heuer with the exception

It appears to the authors of the present study, however, that the higher estimates, which include a percentage of specialists with higher education in the science sector, have a greater comparability to the NSF R&D employment series because they capture a large number of individuals with higher education but without advanced degrees who are engaged in activities that would be considered development in the United States, but who are excluded from the Soviet scientific workers series.

Depending on which estimate is used, the number of Soviet full-time equivalent scientists and engineers engaged in R&D was from one-and-one-third to more than two-and-one-fourth times the number in the United States in 1976. In 1960, on the other hand, all three estimates show the United States as having a greater number of full-time equivalent scientists and engineers engaged in R&D, with a range of from 16.1 percent greater than the highest estimate for the Soviet Union to 62.1 percent greater than the lowest estimate. The average annual rate of growth of U.S. full-time equivalent scientists and engineers engaged in R&D from 1961 to 1976 was 2.3 percent. By all three estimates for the Soviet Union contained in Table VII-1, the average annual rate of growth has been considerably higher than that for the United States, with a 7.6 percent average annual growth rate for estimates in Column A, 8.6 percent for estimates in Column B[1] and 8.7 percent for estimates in Column B[2].

Campbell has also prepared estimates of Soviet expenditures on R&D as a percent of GNP. (A discussion of definitions and concepts underlying U.S. and Soviet statistics on research and development is contained in Appendix B.) These estimates have been used by the National Science Foundation for comparison with U.S. R&D expenditures, as shown in Table VII-2. Although Campbell cautions that the Soviet estimates may be significantly inflated compared to U.S. data,[7] the estimates provide some indication of relative trends in emphasis on performance of R&D in the two economies. Since 1964, U.S. expenditures on R&D as a percent of GNP have declined from 2.97 percent to 2.25 percent in 1978, a percentage drop of about 24 percent. By contrast, the Soviet Union appears to have been increasing the emphasis on performance of R&D in the economy. From 1963 to 1977, the United States showed an average annual decline in expenditures

---

that geography is also included. In their estimates which exclude only the humanities, economics, psychology, law and education are added to the above. Full-time equivalent graduate student participation in R&D is estimated as 25 percent of full-time aspirants and 10.5 percent of part-time aspirants.

[7]Campbell, op. cit., p. 15.

TABLE VII-2

U.S. AND U.S.S.R. EXPENDITURES FOR PERFORMANCE OF R&D
AS A PERCENT OF GROSS NATIONAL PRODUCT (GNP): 1962-78

YEAR	U.S.	U.S.S.R.
1962	2.73	2.64
1963	2.87	2.80
1964	2.97	2.87
1965	2.91	2.85
1966	2.90	2.88
1967	2.91	2.91
1968	2.83	NA
1969	2.74	3.03
1970	2.64	3.23
1971	2.50	3.29
1972	2.43	3.58
1973	2.34	3.66
1974	2.32	3.64
1975	2.30	3.69
1976	2.27	3.55
1977	2.27	3.47
1978	2.25	NA

Source: National Science Board, Science Indicators 1978, NSB 79-1,
         p. 140 (1979).

on R&D as a percent of GNP of −1.2 percent, while the Soviet Union showed an average annual rate of growth of 1.8 percent.

B. Soviet Scientific Workers by R&D Employment Sector

As noted above, organizations that perform R&D in the Soviet Union fall into the following four sectors: academies of sciences institutes, scientific organizations of higher educational institutions, branch scientific organizations, and research units of industrial enterprises. As shown in Table VII-3, there were 5,327 scientific establishments, including higher educational establishments, in the USSR in 1975, a 27 percent increase over the number in 1960.

The distribution of Soviet scientific workers by sector of employment from 1960 to 1973 is shown in Table VII-4. In 1973, 8.8 percent of all scientific workers were employed in the academy of sciences system, 35.6 percent at higher educational institutions, 46.6 percent in branch scientific organizations, and 9.0 percent in research units of industrial enterprises.

The percentage of scientific workers employed in academy of sciences institutes and higher educational institutions had declined since 1960, with 8.8 percent of total scientific workers employed in the academy system in 1973 as opposed to 17.8 percent in 1960, and 35.6 percent in higher educational institutions in 1973 as opposed to 41.5 percent in 1960. Percentage employment both in branch scientific organizations and in research units of industrial enterprises, on the other hand, had increased during the period, rising from 38.8 percent to 46.6 percent of total scientific workers in the case of branch institutes, and from 2.0 percent to 9.0 percent of the total in the case of industrial enterprise research units.

The basic structure and characteristics of each of the four main categories of R&D performer sectors in the Soviet Union and the scientific workers employed in each sector are discussed below.

1. Academies of Sciences Institutes

Scientific organizations within the network of the academies of sciences include those of the Academy of Sciences of the USSR, the academies of sciences of the union republics, and the specialized branch academies under the supervision of the Academy of Sciences of the USSR but subordinate to various ministries. Branch academies, which were established for the purpose of consolidating the activities of academy institutions in important areas of national economic activity, currently

TABLE VII-3

U.S.S.R. SCIENTIFIC ESTABLISHMENTS: 1940 to 1975

	1940	1950	1960	1965	1970	1975
Total (including higher educational institutions)	2,359	3,447	4,196	4,867	5,182	5,327
Scientific research institutes, including their affiliates and departments	786	1,157	1,728	2,146	2,525	2,805
Higher educational institutions	817	880	739	756	805	856

SOURCES: The National Economy of the U.S.S.R. in 1975, p. 681;
          The National Economy of the U.S.S.R. in 1974, pp. 143 and 691;
          The National Economy of the U.S.S.R. in 1968, p. 694; and
          The U.S.S.R. in Numbers in 1975, pp. 69 and 227

TABLE VII-4

U.S.S.R SCIENTIFIC WORKERS, BY EMPLOYMENT SECTOR: 1960 TO 1973

	1960		1965		1970		1973	
	Thousands	Percent	Thousands	Percent	Thousands	Percent	Thousands	Percent
Total scientific workers	354.2	100.0	664.6	100.0	927.7	100.0	1,108.5	100.0
In academy of sciences institutes	62.9	17.8	61.3	9.2	85.9	9.3	97.0	8.8
In higher educational institutes	146.9	41.5	221.8	33.4	348.8	37.6	394.4	35.6
In branch scientific organizations	137.2	38.8	329.1	49.5	419.1	45.2	516.8	46.6
In research units of industrial enterprises	7.2	2.0	52.4	7.9	73.9	8.0	100.3	9.0

SOURCE: Louvan E. Nolting and Murray Feshbach, "R&D Employment in the USSR - Definitions, Statistics and Comparisons" in The Soviet Economy in a Time of Change, Vol. I, Joint Economic Committee, U.S. Congress, p. 728 (October 1979).

exist in the fields of agriculture, medicine, education, the humanities, and municipal economy.

The scientific research institutions of the academies of sciences are primarily engaged in conducting basic research in the natural and social sciences. It has been estimated that between 65 and 80 percent of basic research in the Soviet Union is conducted by the academy network.[8] The majority of institutes within the system of the academies of sciences, however, conduct some amount of applied research. Results of a selective survey of scientific workers in the Academy of Sciences indicated that 67.8 percent of the senior scientific workers and Doctors of Science and 47 percent of the Candidates of Science in the Academy were engaged in basic research, while the percentage of such specialists engaged in applied research was 8.8 percent and 38.3 percent, respectively. The remainder were engaged in research of a fundamental-applied character.[9] Another study conducted in 1974 showed the following average division of time spent on various types of research by scientists within the Academy of Sciences: 57 percent on basic research, 30 percent on applied research, 8 percent on development, and 5 percent on introduction of the results of scientific research into production.[10] While most institutes of the academies of sciences conduct some applied research, there are almost no institutes that are engaged primarily in applied research, except for a few establishments whose principal function is one of servicing the main research institutes of the academies of sciences. The latter include, for example, the All-Union Institute for Scientific and Technical Information, the Institute for Information on Social Sciences, and the Design Bureau for Scientific Instrument Engineering.

The Academy of Sciences of the USSR is not analogous to the National Academy of Sciences of the United States. In contrast with the Soviet Academy, the U.S. National Academy is an autonomous honorary society which, although supported largely by government funds, is not a government agency, has no system of research institutes within its organizational makeup, and has no formal role in government science and technology policy decision-making. The Academy of Sciences of the USSR is considered the most prestigious scientific or-

[8]Louvan E. Nolting, "The Financing of Research, Development, and Innovation in the U.S.S.R. by Type of Performer," *Foreign Economic Report No. 9*, U.S. Department of Commerce, p. 47 (1965).

[9]S.A. Kugel' et al., "Answer to Questions of the American Side on the Survey Report 'Training and Utilization of Scientific and Engineering-Technical Personnel in the USSR: Part II'," Moscow, p. 5 (December 1977).

[10]S.R. Mikulinskiy et al., "The Training and Utilization of Scientific and Engineering-Technical Personnel in the USSR: Part II," Moscow, p. 38 (December 1976).

ganization in the Soviet Union, employing many of the most highly qualified scientists. The Soviet Academy is responsible for coordinating all fundamental research in the natural and social sciences conducted throughout the country.

Twenty years ago, the overwhelming majority of both the scientific institutions and the scientists of the Soviet Academy of Sciences were concentrated in Moscow and Leningrad. However, with the increasing need for basic research directed toward the solution of economic problems of the regions of the country remote from the center, and for the training of the scientifically gifted youth of those regions, a considerable expansion of the locations of activity of the Academy took place. At the end of the 1950s, a Siberian department of the Academy was created, which is now the largest scientific center in Siberia. The Siberian department now includes more than 20 percent of all scientific institutions of the Academy, and more than 15 percent of all its scientists. Additional centers of the Soviet Academy of Sciences have been created in the Far East and the Urals, and branches have been established in Bashkiria, Tataria, Dagestan, Karelia, the Komi Autonomous Republic, and the Kola peninsula. In almost all of these regional scientific centers and branches, the qualification level of the scientific personnel differs little from the current average qualification level in the Academy as a whole.

The Soviet Academy of Sciences directs and coordinates the activity of the academies of sciences of the individual republics. In the latter, in addition to basic research in the natural and social sciences, there is some emphasis on applied research. Table VII-5 shows the number of scientific institutions and scientific workers associated with the Soviet Academy of Sciences, the academies of sciences of the various republics, and the branch academies at the end of 1975. The academies of the individual republics included about 400 scientific institutions, more than 100 of which had been established since 1960. In 1975, about 45,400 scientists were employed by the republic academies, 5.4 percent of whom were Doctors of Science and 39.3 percent of whom were Candidates of Science.

The scientific research subdivisions of the branch academies include a considerable proportion of the specialists of that particular scientific discipline. For example, the research institutes of the All-Union Academy of Agricultural Sciences employ about 25 percent of Soviet specialists in agricultural sciences, the Academy of Medical Sciences employs about 8 percent of the medical specialists, and the Academy

TABLE VII-5

SOVIET ACADEMY OF SCIENCES, ACADEMIES OF SCIENCES OF THE UNION REPUBLICS, AND BRANCH ACADEMIES: END OF 1975

	Year Established	Active and Associate Members	Scientific Establishments Which Belong to the Academy	Scientific Workers	Doctors of Science	Candidates of Science
Academy of Sciences of the USSR	1724[1]	678	246	41,836	3,633	18,553
Academy of Sciences of Ukrainian SSR	1919	282	76	12,102	822	5,465
Academy of Sciences of Belorussian SSR	1928	131	33	4,640	173	1,263
Academy of Sciences of Uzbek SSR	1943	96	31	3,699	172	1,524
Academy of Sciences of Kazakh SSR	1945	132	33	3,731	177	1,493
Academy of Sciences of Georgian SSR	1941	109	40	5,493	332	1,733
Academy of Sciences of Azerbaydzhan SSR	1945	90	32	4,222	244	1,685
Academy of Sciences of Lithuanian SSR	1941	39	12	1,534	53	707
Academy of Sciences of Moldavian SSR	1961	37	19	883	56	492
Academy of Sciences of Latvian SSR	1946	52	16	1,760	68	702
Academy of Sciences of Kirghiz SSR	1954	44	19	1,434	60	526
Academy of Sciences of Tadzik SSR	1951	42	19	1,213	47	476
Academy of Sciences of Armenian SSR	1943	90	31	2,835	170	898
Academy of Sciences of Turkmen SSR	1951	49	16	866	35	393
Academy of Sciences of Estonian SSR	1946	44	16	949	56	500
Academy of Art USSR	1947	130	5[2]	386	18	134
All Union Order of Lenin Academy of Agricultural Sciences	1929	211	166	10,339	429	5,133
Academy of Medical Sciences of USSR	1944	271	40	5,480	862	3,201
Academy of Pedagogical Sciences of USSR[3]	1943	131	14	1,711	122	749
Academy of the Communal Economy of the RSFSR	1931	--	5	427	10	216

[1] The opening of the Academy took place in 1725.

[2] Including two higher educational institutions.

[3] Until 1956, this was the Academy of Pedagogical Sciences of the RSFSR

Source: The National Economy of the USSR in 1975, p. 167.

of Pedagogical Sciences employs about 5 percent of all education specialists.[11]

Table VII-6 shows the distribution of scientific workers of the Soviet Academy of Sciences by branch of science for the end of 1970. The proportion of scientific workers employed by the Academy compared to all scientific workers in the USSR varied for different branches of science. In such branches as chemistry and biology, it exceeded 12 percent; in physics, mathematics and history, it was about 9 percent; while in the technical sciences, education and medicine, it was 1 percent or less. The relatively small proportion of scientific workers in the education and technical sciences branches is explained by the existence of the branch education academies and by the fact that in the 1960s those sections of the Academy that conducted research in the technical sciences were transferred to branch ministries and departments.

More than half of all scientific workers of the Soviet Academy of Sciences were physicists, mathematicians, chemists, and biologists. Social scientists comprised more than a fifth of the scientists of the Academy, with approximately 7 percent in economics and 4 percent in philology.

2. Scientific Organizations of Higher Educational Institutions

Higher educational organizations in the USSR account for only about 5 percent of total Soviet R&D and 10 to 13 percent of basic research.[12] This is in sharp contrast to the United States, where universities and colleges conduct about 13 percent of total R&D and 60 percent of basic research.[13] However, the amount of research performed in Soviet higher educational organizations has been steadily increasing (between 1965 and 1975, for example, it quadrupled), and is likely to become an even more significant component of the Soviet R&D effort as measures are taken to make greater utilization of the large number of scientific personnel employed in this sector.

The principal organizational units responsible for the conduct of research at Soviet higher educational institutions are the scientific divisions and laboratories of the academic departments (kafedry), scientific research and design institutes, problem laboratories, branch

[11]S.R. Mikulinsky et al., "The Training and Utilization of Scientific and Engineering-Technical Personnel in the USSR: Part II," Moscow, p. 89 (December 1976).
[12]Nolting, "The Financing of Research, Development, and Innovation in the U.S.S.R. by Type of Performer," op. cit., p. 47.
[13]Including associated federally funded research and development centers. *National Patterns of R&D Resources 1953-1978-79*, National Science Foundation, NSF 78-313, p. 4 (October 1978).

TABLE VII-6

DISTRIBUTION OF SCIENTIFIC WORKERS OF THE SOVIET
ACADEMY OF SCIENCES, BY BRANCH OF SCIENCE: 1970
(In Percent)

	Scientific Workers in the Soviet Academy of Sciences as a Percent of Total Scientific Workers	Distribution of Scientific Workers in the Soviet Academy of Sciences by Branch
Total scientific workers	3.7	100.0
Physics and mathematics	9.2	25.2
Chemistry	12.3	16.1
Biology	12.6	13.6
Geology and mineralogy	12.1	7.1
Technical sciences	1.0	11.6
Agriculture	1.9	1.7
History	9.9	7.2
Economics	4.3	7.1
Philosophy	4.0	1.4
Philology	3.0	4.2
Geography	11.5	2.4
Law	5.3	0.7
Education	1.0	0.9
Medicine	0.4	0.5
Art	0.4	0.1
Psychology	1.4	0.1

Source: Gvishiani, et. al., The Scientific Intelligentsia in the U.S.S.R.,
Moscow, Progress Publishers, p. 180 (1976).

laboratories, and independent subdivisions of higher educational institutions called "scientific research sectors." Most of the research
conducted by scientific subdivisions of academic departments consists
of basic research. Problem laboratories primarily conduct fundamental
research on major national problems of interest to the academic departments with which they are associated or as specified by the Soviet
Academy of Sciences. Branch laboratories primarily conduct applied
research under contract with ministries or industrial enterprises. Scientific research sectors in smaller higher educational institutions are
often involved in all research activities conducted in their particular
institution, while in larger higher educational institutions they may be
responsible only for contractual research. In 1975, there were more
than 60 scientific research institutes and design bureaus, more than
1,000 problem laboratories and branch laboratories, and about 300
scientific research sectors associated with Soviet higher educational
institutions.

Prior to the mid-1950s, more than half of the scientific workers in
the USSR were employed in higher educational institutions. Since the
latter part of the 1950s, however, the proportion has been declining
so that by the mid-1970s only about 35 percent of all scientific workers
in the USSR were employed in higher educational institutions. (See
Table VII-4, above) However, except for a few selected fields, while
higher educational institutions no longer employ the majority of scientific workers in the USSR, the number of scientific pedagogical
personnel who are engaged full-time in R&D has been increasing
sharply. The distribution of scientific-pedagogical workers by position
held from 1955 to 1974 is shown in Table VII-7. While the total
number of scientific workers employed in higher educational institutions has more than tripled since 1955, the most dramatic increase in
workers holding various types of positions has been in the number of
scientific workers engaged full-time in R&D (scientific workers who
do not teach), which by 1974 was more than 25 times the number in
1955.

As shown in Table VII-8, in 1955 only 2.2 percent of scientific-
pedagogical personnel were engaged full-time in R&D, while by 1976
the proportion had risen to 17.9 percent. This has been attributed to
an increased emphasis of the regime on utilizing scientific-pedagogical
personnel for performing applied R&D under contract with industrial
enterprises. In 1976, between 80 and 90 percent of the R&D performed
by Soviet higher educational institutions was contractual research.[14]

[14]Heuer, op. cit., p. 36.

TABLE VII-7

U.S.S.R. SCIENTIFIC-PEDAGOGICAL WORKERS,
BY POSITION HELD: 1955-1975
(In Thousands)

	1955	1960	1970	1974
Number of scientific-pedagogical personnel	119,059	146,915	348,872	410,818
Doctor of Science	5,547	5,967	11,634	15,283
Candidate of Science	43,450	51,911	111,094	145,793
Rector and assistants for teaching and scientific work	1,877	2,057	2,890	3,224
Doctor of Science	365	362	586	797
Candidate of Science	980	1,223	1,745	1,854
Deans of departments	2,262	2,876	4,878	5,180
Doctor of Science	238	229	327	468
Candidate of Science	1,369	1,836	3,095	3,800
Heads of departments	15,114	16,901	25,298	28,829
Doctor of Science	3,783	4,136	6,555	8,279
Candidate of Science	7,693	9,204	13,962	15,735
Professors in composition of departments	1,437	1,547	4,643	6,124
Doctors of Science	920	1,022	3,346	4,641
Candidate of Science	139	185	835	1,015
Docents in composition of departments	19,036	25,940	60,018	73,894
Doctor of Science	122	121	345	402
Candidate of Science	16,589	23,048	54,072	68,927
Senior instructors, instructors and assistants	76,354	91,817	205,528	223,450
Doctor of Science	93	124	207	233
Candidate of Science	16,185	15,301	30,336	42,656
Scientific workers who do not teach	2,678	5,777	45,617	70,496
Doctor of Science	26	34	268	463
Candidate of Science	787	1,114	7,049	11,806

Note: Figures reported by Zhiltsov in the Soviet report appear to be in error;
except for data for 1970, the figures in the various categories do not
equal the reported total. The number of scientific workers who do not
teach for 1960 and 1974 reported by Nolting and Feshbach (see Table VII-8)
was therefore substituted to closer approximate the reported total. In
addition, Zhiltsov appears to have erroneously labeled the data in Column 4
as "1975" figures, when the reported total corresponds to the number of
scientific-pedagogical workers in 1974, which was therefore changed to
that year.

Source: E. Zhiltsov, et. al., "The Training and Utilization of Scientific
Engineering and Technical Personnel in the U.S.S.R., Part I," Faculty of
Economics, Economics of Education Laboratory, Moscow (November 1977).

TABLE VII-8

U.S.S.R. SCIENTIFIC PEDAGOGICAL WORKERS
ENGAGED FULL-TIME IN R&D:
1950-1976

Year	Total Scientific Pedagogical Workers	Scientific-Pedagogical Workers Engaged Full-Time in R&D	Percent
1950	86,542	2,070	2.4
1955	119,059	2,678	2.2
1960	146,915	5,777	3.9
1965	221,800	17,000	7.7
1970	348,872	45,617	13.1
1974	410,818	70,496	17.2
1976	441,500	79,200	17.9

Source: Louvan E. Nolting and Murray Feshbach, "R&D Employment in the USSR--
Definitions, Statistics, and Comparisons," in Soviet Economy in a
Time of Change, Vol. I, Joint Economic Committee, U.S. Congress,
p. 733 (October 1979).

Soviet educational specialists have sometimes complained that the increase in contract work that is basically of an applied nature may detract from the need for the conduct of fundamental research, which is thought to be more conducive to the overall educational goal of training scientists. Other specialists, however, have emphasized that the increase in contract work has the advantage of endowing research with a practical purpose and of strengthening the link between teaching personnel and industrial enterprises. Along these lines, a decree on higher education issued in July 1979 called for measures to improve the organization of research at higher educational institutions and deplored the delay in measures to equip higher educational instititions with modern laboratories and scientific equipment.

A particular advantage that has been noted regarding the conduct of research at higher educational institutions is that because of the availability of specialists from a wide range of fields, higher educational institutions are particularly suitable for conducting research of an interdisciplinary nature. However, this does not appear to have yet taken place on a very wide scale. A selected study of higher educational institutions in Leningrad found that only about one out of every ten to twelve research projects is conducted on an interdepartmental basis.[15]

The proportion of scientific workers who were employed in higher educational institutions, by branch of science, in 1970 is shown in Table VII-9. Higher educational institutions still employed over 60 percent of scientific workers in the social sciences and humanities as a whole, and, except for economics and geography, employed well over 50 percent of all scientific workers in each specific field in the social sciences and humanities. By contrast, only 22.5 percent of scientific workers in the technical sciences (engineering) were employed in higher educational institutions.

Various estimates have been made of Soviet graduate student participation in R&D conducted by higher educational institutions or by the research organizations at which they undergo their training. While the Doctor of Science degree in the Soviet Union is generally awarded only after the successful defense of a doctoral dissertation representing original research involving the solution of a problem of major significance to the development of science, because the doctoral degree involves no prescribed academic program but is usually awarded to mature scientists, most of whom already hold the Candidate of Science degree, individuals seeking the Doctor of Science degree are probably almost all included in the statistics on scientific workers. While data

[15]Gvishiani et al., op. cit., p. 182.

TABLE VII-9

U.S.S.R. SCIENTIFIC-PEDAGOGICAL WORKERS AS A PERCENT OF TOTAL SCIENTIFIC WORKERS,
BY BRANCH OF SCIENCE: 1970

	Number of Total Scientific Workers	Number of Scientific-Pedagogical Workers	Scientific-Pedagogical Workers as a Percent of Scientific Workers
Physical & Life Sciences	284,174	113,160	39.8
Physics/Mathematics	95,272	46,499	48.8
Chemistry	45,815	15,222	33.2
Biology	37,342	11,196	30.0
Geology/Mineralogy	20,342	3,779	18.6
Agriculture/Veterinary Sciences	35,446	9,862	27.8
Medicine/Pharmacy	49,957	26,602	53.2
Engineering (Technical Sciences)	409,470	92,058	22.5
Social Sciences & Humanities	200,812	124,976	62.2
History and Philosophy	37,177	23,178	62.3
Economics	57,518	22,012	38.3
Philology	48,721	41,476	85.1
Geography	7,242	2,871	39.6
Law	4,765	2,748	57.7
Education	31,283	22,507	71.9
Art	12,182	8,956	73.5
Psychology	1,924	1,228	63.8
Architecture	2,590	1,320	51.0
Other	30,663	17,358	56.6
Total	927,709	348,872	37.6

Source: B. M. Remennikov, The Higher School in the System of Reproduction of the Labor Force in the USSR,
Moscow, p. 155 (1973).

are not available on the number of persons who have completed their aspirant training and are conducting research in preparation for the Candidate dissertation, these individuals, if employed, are also likely to be included in the scientific workers series. Thus, to account for full-time equivalent participation in R&D by Soviet graduate students, calculations are generally based on the number of full-time and part-time students enrolled in formal aspirant programs.

One Soviet source has estimated that graduate students devote approximately one-fourth of their time to research.[16] However, other Soviet commentators have stressed that students enrolled in full-time aspirant programs must devote so much of their time to attending courses and seminars that they have little time for research or preparation of dissertations. For this reason, Western analysts use estimates ranging from as little as 10 percent of full-time aspirants[17] to as high as 25 percent of full-time aspirants plus 10.5 percent of part-time aspirants.[18] The number of aspirants, reduced to full-time equivalent scientists and engineers engaged in R&D on the above bases, from 1960 to 1978, is shown in Table VII-10.

3. Branch Scientific Organizations

Branch scientific organizations are subordinate to the USSR and union republic ministries and departments, and encompass a variety of generalized and specialized institutions that conduct scientific research relating to their respective sectors of the economy. The activities of branch scientific organizations range across the entire research to production cycle and include some basic research, although the majority of their research work is of an applied nature. Branch scientific research institutes conduct a major portion of the applied research (approximately 85 percent)[19] and a considerable amount of the development work of the Soviet Union.

Table VII-11 shows the number of scientific personnel employed in branch scientific organizations from 1960 to 1973. From 1960 to 1973, the number of scientific personnel in branch scientific research

---

[16]Nolting and Feshbach, op. cit., p. 734, citing Feshbach "Notes on R&D Manpower" Based on Discussions in Moscow and Leningrad, June 26-30, p. 9 (1978).

[17]Campbell, op. cit., p. 42.

[18]Nolting and Feshbach, op. cit., p. 734. rather than, The Defense Intelligence Agency's series of reports on "Soviet Professional, Scientific, and Technical Manpower" uses even higher estimates based on 50 percent of full-time aspirants; this approach, however, is probably too generous.

[19]Nolting, "The Financing of Research, Development, and Innovation in the U.S.S.R. by Type of Performer," op. cit., p. 47.

TABLE VII-10

U.S.S.R. GRADUATE STUDENT PARTICIPATION IN R&D: 1960 to 1974
(In absolute numbers, except for percent)

		1960	1965	1970	1974
1.	Total numbers of graduate students	36,754	90,294	99,427	96,939
2.	Percent of total graduate students in science and engineering	77.3	80.4	76.8	74.8
3.	Number of full-time students	22,978	51,109	55,024	45,357
4.	Number of part-time students	13,776	39,185	44,403	51,582
5.	Full-time students in science and engineering	17,762	41,092	42,258	33,927
6.	Part-time students in science and engineering	10,649	31,505	34,102	38,583
7.	Number of graduate students as full-time equivalent scientists and engineers in R&D:				
	Estimate A (10% of line 5)	1,776	4,109	4,226	3,393
	Estimate B (25% of line 5 plus 10.5% of line 6)	5,559	13,581	14,145	12,533

Note: Methodology for Estimate A suggested by Campbell, op. cit., pp. 41-42;
       for Estimate B, by Nolting and Feshbach, op. cit., p. 734.

Sources: Lines 1 and 2: The National Economy of the U.S.S.R. in 1960, p. 789;
                          The National Economy of the U.S.S.R. in 1965, p. 716;
                          The National Economy of the U.S.S.R. in 1970, p. 662;
                          The National Economy of the U.S.S.R. in 1974, p. 148.
          Lines 3 and 4: The National Economy of the U.S.S.R. in 1974, p. 147;
                          The National Economy of the U.S.S.R. in 1970, p. 661.

TABLE VII-11

U.S.S.R. SCIENTIFIC WORKERS IN BRANCH SCIENTIFIC INSTITUTIONS:
1960-1973

	SCIENTIFIC WORKERS IN BRANCH SCIENTIFIC INSTITUTIONS (In Thousands)	AS A PERCENT OF TOTAL U.S.S.R. SCIENTIFIC WORKERS
1960	137.2	38.7
1961	184.6	45.7
1962	244.1	46.5
1963	273.1	48.3
1964	300.1	49.0
1965	329.1	49.5
1966	330.6	46.4
1967	355.9	46.2
1968	377.3	45.9
1969	407.1	46.1
1970	419.1	45.2
1971	463.0	46.2
1972	490.6	46.5
1973	516.8	46.6

SOURCE: Louvan E. Nolting and Murray Feshbach, "R&D Employment
in the U.S.S.R.--Definitions Statistics and Comparisons,"
in The Soviet Economy in a Time of Change, Vol. I,
Joint Economic Committee, U.S. Congress, p. 728
(October 1979).

institutes more than tripled, with the greatest increase occurring between 1960 and 1965, during which time the number more than doubled. In 1965, scientific workers in the branch institutes comprised almost half of the total number of scientific personnel in the USSR. During the years since 1965, the rates of increase in scientific personnel employed at branch scientific research institutes decreased somewhat, and the proportion so employed compared to scientific personnel in the USSR as a whole has stabilized at the level of about 46 percent.

An explanation that has been provided for the Soviet regime's increased emphasis on branch research is that unlike basic research, applied research and development have a well-defined direction, dealing with the resolution of problems that arise concerning production and technology in a specific branch of material production. Because the effectiveness of such work largely depends on extensive interaction between the organization performing such research and the appropriate industries, the high degree of centralization of the Soviet Academy of Sciences was regarded as unsuitable by the Soviet regime for the organization of such activities. In addition, the planned system of the economy and the absence of competition made it possible to concentrate a significant part of scientific personnel in scientific research institutes directly subordinate to the various ministries and departments. That each ministry confirms the plan for scientific R&D within its particular economic sector, administers funding provided from the state budget, and manages the R&D activities of its subsidiary scientific research organizations, represented another factor that made it possible to coordinate scientific and technical policy and activities along branch lines.[20]

A relatively new and increasingly important subcomponent of branch scientific organizations is that of scientific production associations, which have been established so as to combine within a single unit the entire research to production cycle. Scientific production associations are organizations involving a combination of research institutes, design bureaus, technological organizations, experimental plants or bases, and production plants within a single administrative unit, usually under the leadership of a scientific research institute. The distinguishing feature of these associations is the inclusion within their structure of a plant which can be used for the first serial production of new products or technologies.

[20]S.R. Mikulinskiy et al., "The Training and Utilization of Scientific and Engineering-Technical Personnel in the USSR: Part II," Moscow, p. 89 (December 1976).

4. Research Units of Industrial Associations and Enterprises

Another role in the research to production cycle in the Soviet Union is provided by scientific research, design, and technological subdivisions of industrial associations and enterprises, sometimes referred to as the "plant" sector or "factory science." Much of the work of enterprise subdivisions consists of practical adaptation of R&D conducted by other scientific organizations and the development and testing of prototypes. Laboratories of the "plant" sector (factory laboratories) are found in three basic organizational forms: central plant laboratories, which sometimes include a series of scientific and technical subdivisions; shop section laboratories, which conduct research under the supervision of the central plant laboratory; and laboratories subordinate to the factory management division. In addition to plant laboratories, the factory science sector includes various design bureaus and experimental facilities that are involved to some extent in the conduct of R&D. Enterprise scientific subdivisions are estimated to account for approximately 5 to 6 percent of total Soviet R&D.[21]

In 1967, there were more than 16,000 factory laboratories, including approximately 9,000 central plant laboratories, in the Russian Republic alone. (See Table VII-12.) Although factory laboratories are the basic specialized scientific subdivisions of industrial enterprises, such enterprises usually also employ some scientific workers in their design and experimental subdivisions or in management.

Scientific workers are utilized not only by industrial enterprises, but also by the design and drafting organizations subordinate to ministries and departments. A considerable percentage of all scientific workers are employed in such organizations. For example, in 1971, 11 percent of scientists in the chemical industry were employed in design and drafting organizations, while 12 percent were employed by industrial enterprises.

In the last two decades, there has been a high rate of increase in the number of scientific workers employed in industrial enterprises, design and drafting organizations, and management. (See Table VII-13.) In 1950, the number of scientific workers so employed was only slightly over 5,500 (3.4 percent of all scientific workers in the USSR). By 1973, this number had increased more than 18 times to include 100,300 persons (9.1 percent of all scientific workers in the USSR).

---

[21]Nolting, "The Financing of Research, Development, and Innovation in the U.S.S.R. by Type of Performer," op. cit., p. 47.

TABLE VII-12

NUMBER OF LABORATORIES, DESIGN AND EXPERIMENTAL
ORGANIZATIONS IN INDUSTRIAL ENTERPRISES IN THE RSFSR: 1967

	Number	Percent
Laboratories	16,011	60.2
Central (plant, factory)	8,844	33.4
Workshop	5,026	18.9
In factory management departments	2,101	7.9
Design Organizations	8,123	30.6
Independent design organizations financed by enterprises	1,902	7.2
Department of the chief designer design offices, sectors and groups in departments of factory management and workshops of enterprises	6,221	23.4
Experimental Organizations	1,726	6.5
Departments of Mechanization and Automation	721	2.7
TOTAL	26,581	100.0

SOURCE: The National Economy of the U.S.S.R. in 1967, Moscow
"Statistika," p. 69 (1968).

TABLE VII-13

U.S.S.R. SCIENTIFIC WORKERS IN INDUSTRIAL ENTERPRISES:
1950 TO 1973

Year	Number of Scientific Workers in Industrial Enterprises (In Thousands)	As a percent of Total U.S.S.R. Scientific Workers
1950	5.5	3.4
1955	8.3	3.7
1960	7.2	2.0
1965	52.4	7.9
1970	73.9	8.0
1971	82.8	8.3
1972	92.0	8.7
1973	100.3	9.0

Source: Louvan E. Nolting and Murray Feshbach, "R&D Employment in
the USSR--Definitions, Statistics, and Comparisons,"
in The Soviet Economy in a Time of Change, U.S. Congress,
p. 728 (October 1979).

The following is a list of branches of industry in terms of the percentage of their total employees who are scientific workers, given in decreasing order:[22]

Machine-building and metalworking
Chemical and petroleum
Nonferrous metallurgy
Ferrous metallurgy
Glass and earthenware
Fuel
Light
Food
Wood, woodworking, and pulp and paper
Construction material

The research units subsidiary to industrial enterprises are generally small, averaging about 10 scientific workers. However, in the ferrous metallurgy, metallurgical machine-building, and instrument manufacture industries, the number of scientific workers per scientific research unit is almost double the average for industry as a whole, while in the light and food industries it is only about half the average.[23]

C. U.S. Scientists and Engineers Engaged in R&D, by Sector

In the United States, the diversified organizations that perform R&D are classified in the NSF statistics into the following sectors: Federal government, industry, universities and colleges, federally funded research and development centers (FFRDCs), other nonprofit institutions, state and local government, and foreign performers.

Most of these categories are self-explanatory with regard to the nature and the scope of the R&D organizations they encompass. "Industry" is defined by NSF as "organizations that may legally distribute net earnings to individuals or to organizations." The industry sector is defined principally in contrast to nonprofit institutions, which are private institutions chartered to perform public service, no part of whose net earnings may be distributed to a private stockholder or individual. The term "industry" as used in the NSF classifications includes, in addition to manufacturing proper, some nonmanufacturing

---

[22]Gvishiani et al., op. cit., p. 190.
[23]Gvishiani et al., op. cit., p. 191.

firms such as public utilities, mining, agriculture, forestry, and fisheries.

Federally funded research and development centers are R&D performing organizations exclusively or substantially financed by the federal government in order either to meet a particular R&D objective or, in some instances, to provide major facilities at universities for research and associated training purposes. Some FFRDCs are administered by private industrial firms, some by universities, and some by other nonprofit institutions.

The coverage of the state and local government category is specifically restricted to exclude state or locally supported universities and colleges, agricultural experiment stations, and medical schools and affiliated hospitals, all of which are treated as part of the university and college sector. Organizations are classified in the "universities and colleges" sector if their fundamental purpose is instruction or training above the secondary school level, without regard to whether or not they are profit-making institutions.

When data are aggregated on a national basis, the National Science Foundation uses the following four-sector divisions: Federal government, industry, universities and colleges, and other nonprofit institutions. FFRDCs, with the exception of those administered by universities and colleges, are not reported separately but as part of the sector administering them. Because of the relatively small amount of state and local investment in or performance of R&D other than that treated as part of the universities and colleges sector, as noted above, it is not reported as a separate sector when data are aggregated on a national basis, but as part of the "other nonprofit institutions" sector.

Figure VII-1 shows the distribution of total funds used for performance of research and development in the United States, by R&D performer and by source of funds, for 1980. As illustrated by the figure, most of the funds for R&D originate in the federal government (48.7 percent) and in industry (47.6 percent). Industry, on the other hand, is the dominant performer of R&D, accounting for more than two-thirds of the total, followed by research laboratories of government agencies, with about 13 percent of the total, and universities and colleges, with about 10 percent. Other nonprofit institutions represent a relatively insignificant portion of the total, both in terms of source of funds and in terms of performance of R&D.

The distribution of full-time equivalent scientists and engineers employed in R&D by sector from 1954 to 1980 is shown in Table VII-14. As with the distribution of R&D funds by performer category, industry accounted for more than two-thirds of R&D scientists and

[Dollars in millions] [1]

Sources of funds	Performers						
	Federal Govern- ment	Industry[2]	Univer- sities and colleges[3]	Associated FFRDC's[4]	Other nonprofit institu- tions[2]	Total	Percent distribution, sources
Federal Government	$7,830	$13,950	$4,100	$2,000	$1,520	$29,400	48.7
Industry	——	²28,300	210	——	200	28,710	47.6
Universities and colleges	——	——	⁵1,300	——	——	1,300	2.2
Other nonprofit institutions	——	——	440	——	⁵525	965	1.6
Total	7,830	42,250	6,050	2,000	2,245	60,375	
			8.050				
Percent distribution, performers	13.0	70.0	10.0	3.3	3.7		100.0
			13.3				

[1] All data are estimated from reports by performers.

[2] Expenditures for federally funded research and development centers (FFRDC's) administered by both industry and by nonprofit institutions are included in the totals of their respective sectors. FFRDC's are organizations exclusively or substantially financed by the Federal Government to meet a particular requirement or to provide major facilities for research and training purposes.

[3] Includes agricultural experiment stations.

[4] Federally funded research and development centers (FFRDC's) administered by individual universities and colleges and by university consortia.

[5] Includes State and local government funds.

Source: National Science Foundation, National Patterns of Science and Technology Resources, 1980, NSF 80-308, p. 14.

Figure VII-1   U.S. INTERSECTORAL TRANSFERS OF FUNDS FOR PERFORMANCE OF RESEARCH AND DEVELOPMENT: 1980 (estimated)

TABLE VII-14

U.S. FULL-TIME EQUIVALENT SCIENTISTS AND ENGINEERS EMPLOYED IN R&D,
BY SECTOR: 1954 TO 1980

	1954		1961		1965		1970		1975		1978		1980[1]	
	Thousands	Percent of Total	Thousands	Percent of Total	Thousands	Percent of Total	Thousands	Percent of Total	Thousands	Percent of Total	Thousands	Percent of Total	Thousands	Percent of Total
Total	237.1	100.0	425.7	100.0	494.5	100.0	546.5	100.0	534.9	100.0	601.6	100.0	659.0	100.0
Federal government	37.7	15.9	51.1	12.0	61.8	12.5	69.8	12.8	63.4	11.9	65.4	10.9	66.5	10.1
Industry	164.1	69.2	312.0	73.3	348.4	70.5	375.5	68.7	363.8	68.0	415.8	69.1	465.0	73.9
Universities and colleges[2]	30.0	12.7	51.5	12.1	64.5	13.0	80.0	14.6	82.9	15.5	92.4	15.4	98.5	14.9
Other non-profit institutions	5.3	2.2	11.1	2.6	19.9	4.0	21.2	3.9	24.8	4.6	28.0	4.7	29.0	4.4

[1] Estimate.

[2] Includes associated FFRDCs.

Source: 1970, National Patterns of R&D Resources 1953-1977, NSF 77-310, p. 34; all other years, National Patterns of Science and Technology Resources, 1980,
NSF 80-308, p. 33.

engineers. The share of total scientists and engineers employed by universities and colleges has generally been rising since 1961, with 12.1 percent of the total in 1961 and 15.4 percent of the total in 1978, then declining to 14.9 percent in 1980. The share of the total accounted for by research laboratories of government agencies, on the other hand, has been declining from 15.9 percent in 1954 to 10.1 percent in 1980. About 4.4 percent of total R&D scientists and engineers were employed by the other nonprofit institutions sector in 1980, as opposed to 2.2 percent of the total in 1954.

As noted above, the industry sector is the largest performer of R&D in the United States, accounting for about two-thirds of the total U.S. R&D effort. Industry is a particularly important contributor to the applied research and development components of the total U.S. R&D effort, accounting for about 60 percent of the applied research and about 85 percent of the development performed in the United States in 1980. On the other hand, it accounted for only about 16 percent of U.S. basic research. Of the total R&D conducted by industry, almost 78 percent consisted of development, about 19 percent was applied research, while only about 3 percent consisted of basic research.[24]

The distribution of U.S. R&D scientists and engineers by industry from 1961 to 1977 is shown in Table VII-15. In 1977 more than 380,000 scientists and engineers were employed in R&D activities in industry. Those industries with the largest numbers of scientists and engineers engaged in R&D activities in 1977 were the electrical equipment, aircraft, machinery, and chemicals industries, which together accounted for more than two-thirds of the R&D employment in industry. The employment of R&D scientists and engineers in industry has been rising steadily from a 1972 low of 349,900. The only industries which did not follow this general pattern were the electrical equipment, motor vehicles, aircraft, and nonmanufacturing industries. The large increase in industrial R&D performers has been attributed to increased spending on energy-related R&D from both government and industry sources.[25]

As in industry, the R&D performed by research laboratories of government agencies primarily consists of applied research and development. In 1978, about 50 percent of the R&D performed in this sector consisted of development, 35 percent was applied research, and 15 percent was basic research. The distribution of civilian R&D sci-

[24]*National Patterns of Science and Technology Resources 1980*, National Science Foundation, NSF 80-308, p. 14-15.
[25]National Science Foundation, *National Patterns of R&D Resources 1953–1978–79*, NSF 78-313, pp. 13, 14 (1978).

TABLE VII-15

U.S. FULL-TIME-EQUIVALENT NUMBER OF R&D
SCIENTISTS AND ENGINEERS BY INDUSTRY: JANUARY 1961-77[1]

(In Thousands)

Industry	1961	1962	1963	1964	1965	1966	1967	1968	1969	1970	1971	1972	1973	1974	1975	1976	1977
Total	312.1	310.8	327.3	340.2	343.6	353.2	367.2	376.7	387.1	384.1	366.8	349.9	356.6	358.2	360.8	364.3	380.4
Chemicals and allied products	37.0	36.5	38.3	35.3	37.9	38.6	36.9	38.9	40.1	40.2	42.9	41.3	41.2	42.2	45.4	44.0	46.8
Petroleum refining and extraction	9.0	9.1	8.9	8.1	8.7	8.9	8.7	9.2	10.0	9.9	9.2	8.3	8.2	8.2	8.4	8.6	8.7
Rubber products	5.5	5.6	5.8	6.0	5.8	5.7	5.8	6.1	6.6	7.4	6.7	6.7	7.5	7.6	8.3	8.6	9.2
Stone, clay, and glass products	3.6	3.7	3.8	3.3	3.5	3.1	3.3	4.1	4.2	4.6	4.3	4.1	4.2	4.5	4.4	5.0	5.0
Primary metals	6.9	6.0	5.2	5.1	5.5	5.5	5.9	5.9	6.2	6.3	6.3	6.0	5.5	5.7	5.5	8.1	8.6
Fabricated metal products	8.6	7.4	6.8	7.0	6.6	6.3	6.3	5.6	6.6	5.9	7.1	6.6	6.7	7.3	7.4	6.8	7.2
Machinery	33.0	31.5	31.4	27.3	29.4	30.5	33.6	37.4	39.8	42.3	42.7	43.7	46.1	50.6	52.3	55.7	56.4
Electrical equipment and communication	79.2	82.3	85.8	89.5	87.8	92.0	98.6	98.4	100.4	100.6	91.8	83.7	85.4	82.5	82.2	80.3	84.0
Motor vehicles and other transportation equipment	19.1	20.8	21.1	23.3	24.1	24.8	25.2	24.3	25.2	25.5	28.2	29.7	29.9	29.2	27.8	27.1	28.7
Aircraft and missiles	78.5	79.4	90.7	101.1	99.2	99.3	100.4	101.1	99.7	92.2	78.2	70.8	72.0	70.5	67.4	66.9	69.6
Professional and scientific instruments	11.1	9.8	9.4	10.8	11.5	12.5	13.0	14.1	15.2	15.0	15.1	15.2	16.2	17.3	17.7	18.8	20.0
Other manufacturing industries	12.9	11.7	11.9	13.6	14.0	14.3	15.4	16.5	17.4	17.6	18.8	18.3	18.5	18.4	19.0	19.8	20.8
Other nonmanufacturing industries	7.5	7.0	8.2	9.8	9.6	11.7	14.1	15.1	15.1	16.3	15.6	15.7	15.3	14.4	14.9	14.6	15.3

[1]Excludes social scientists.

Source: National Science Foundation, National Patterns of R&D Resources, 1953-1978-79, NSF 78-313, p. 13.

entists and engineers in the federal government, by agency and occupation, in January 1977 is shown on Table VII-16. Over half of
these professionals were employed by the Department of Defense.
Except for biological scientists, who were primarily concentrated in
the Department of Agriculture and the Department of Health, Education
and Welfare,[26] the Department of Defense also employed the highest
percentage of civilian R&D scientists employed by government agencies in each specific field.

The principal focus of R&D performed by the universities and colleges sector is basic research, which comprised about 60 percent of
the R&D performed in the sector in 1978. About 28 percent of university R&D consisted of applied research and only about 12 percent
consisted of development work. The number of scientists and engineers
primarily engaged in research and development in U.S. universities
and colleges, by field of specialization from 1965 to 1976, is shown
in Table VII-17. The largest percentage of these professionals are
concentrated in the life sciences, which accounted for about 60 percent
of the total throughout the period. The largest rate of increase has been
in the number of other physical scientists (primarily environmental
scientists), which more than doubled since 1965.

Table VII-18 shows the number of U.S. science and engineering
graduate students engaged part-time in research and development activities, by field of specialization. Table VII-19 shows graduate students engaged part-time in R&D as a percentage of total graduate
student enrollees for 1965 and 1976. Between 1965 and 1976, there
was only a slight increase in the level of graduate science and engineering students in part-time R&D, with 14 percent of the total in
1965 and 15 percent of the total in 1976. The relative involvement in
R&D among fields, however, had changed. In 1965, the highest level
of involvement in part-time R&D work was among graduate students
in the life sciences, with 25 percent of the total. By 1976, the percentage
of graduate students involved in R&D in the life sciences had declined
to 20 percent, while in all other fields there had been an increase.

The data in the tables just discussed were based on surveys of
participation by graduate students in R&D. In estimating full-time
equivalent graduate student participation in R&D, the National Science
Foundation utilizes data on the number of graduate students who held
research assistantships rather than the above data generated by an

---

[26]This Department has since been divided into the Department of Health and Human Resources
and the Department of Education.

TABLE VII-16

DISTRIBUTION OF U.S. CIVILIAN R&D SCIENTISTS AND ENGINEERS
IN THE FEDERAL GOVERNMENT BY AGENCY AND OCCUPATION: JANUARY 1977

Agency	Total	Engi-neers	Physical and environ-mental scientists	Mathema-ticians, statis-ticians, and computer scientists	Biological scientists	Social scientists
Total (in thousands)	48.0	23.0	13.6	3.3	5.9	2.1
	Percent distribution					
Department of Defense	50.8	66.5	39.6	76.7	7.6	33.0
National Aeronautics and Space Administration	12.8	20.9	8.3	4.5	1.0	.6
Department of Agriculture.........	9.3	2.0	7.6	3.0	44.2	12.2
Department of Health, Education, and Welfare	6.6	.5	8.6	2.8	23.2	20.4
Department of the Interior	6.6	2.1	15.6	1.7	7.8	2.4
Department of Commerce.........	4.9	1.9	9.9	3.7	6.8	2.1
Energy Research & Development Administration	2.0	2.6	2.7	.5	.1	.2
Other agencies	6.9	3.5	7.8	7.2	9.4	29.0

NOTE: Excludes uniformed military scientists and engineers and administration of research and development.
Percents may not add to 100 because of rounding.
SOURCE: National Science Foundation, based on data of the U.S. Civil Service Commission.

Source: National Science Foundation, National Patterns of R&D
          Resources, 1953-1978-79, NSF 78-313, p. 12 (1978).

TABLE VII-17

U.S. SCIENTISTS AND ENGINEERS PRIMARILY ENGAGED IN RESEARCH AND
DEVELOPMENT IN UNIVERSITIES AND COLLEGES BY FIELD OF SPECIALIZATION:
JANUARY OF SELECTED YEARS[1]
(In Thousands)

Field of specialization	1965	1969	1971	1973	1974	1975	1976
All fields	40.0	47.1	48.3	46.6	47.4	50.0	51.0
Engineering	4.2	5.0	4.8	5.0	4.9	4.8	4.7
Physical and environmental sciences	5.9	7.0	7.3	7.9	8.3	8.1	8.6
Chemistry	2.3	2.9	2.6	2.8	3.0	2.8	2.9
Physics	2.1	2.4	2.4	2.5	2.5	2.4	2.4
Other	1.5	1.9	2.3	2.7	2.8	2.9	3.2
Mathematics	.9	1.7	1.4	1.3	1.5	1.5	1.6
Life sciences	25.0	28.3	30.4	28.0	28.0	30.6	30.8
Social sciences and psy-cology[2]	4.0	5.2	4.2	4.4	4.7	5.0	5.3

[1] Excludes graduate students.
[2] Excludes history.

NOTE: Components may not add to totals because of independent rounding.

Source: National Science Foundation, National Patterns of R&D
           Resources, 1953-1977, NSF 77-310, p. 15 (1977).

TABLE VII-18

U.S. GRADUATE STUDENTS ENGAGED PART-TIME IN RESEARCH AND DEVELOPMENT
BY FIELD OF SPECIALIZATION: JANUARY OF SELECTED YEARS[1]
(In Thousands)

Field of specialization	1965	1969	1971	1973	1974	1975	1976
All fields	27.2	35.8	37.2	34.6	36.9	39.7	40.2
Engineering	6.4	7.9	8.9	8.4	9.4	11.1	11.0
Physical and environmental sciences	8.1	10.6	10.5	8.8	9.2	9.0	9.2
Mathematics	.9	1.6	1.5	1.5	1.5	1.4	1.3
Life sciences	8.5	10.0	11.2	10.5	10.7	10.8	11.4
Social sciences and psy- cology[2]	3.3	5.7	5.1	5.4	6.2	7.4	7.3

[1] Based on data obtained in the NSF Survey of Scientific and Engineering Personnel Employed at
Universities and Colleges for 1965-74, and the Survey of Graduate Science Student Support and
Postdoctorals for the years 1975-76 See technical notes
[2] Excludes history

NOTE Components may not add to totals because of independent rounding

Source: National Science Foundation, National Patterns of R&D
Resources, 1953-1977, NSF 77-310, p. 15 (1977).

TABLE VII-19

U.S. GRADUATE SCIENCE AND ENGINEERING STUDENTS ENGAGED PART-TIME IN R&D
AS A PERCENT OF TOTAL GRADUATE SCIENCE AND ENGINEERING ENROLLMENT:
1965 AND 1976
(In Percent)

	1965	1976
Engineering	11	19
Physical and Environmental Sciences	22	26
Mathematics	4	5
Life Sciences	25	20
Social Sciences and Psychology	7	9
All Fields	14	15

Source: National Science Foundation, National Patterns of R&D Resources: 1953-1977, NSF 77-310, p. 15 (1977), discussion in text.

employment survey.[27] Table VII-20 shows the number of total graduate students in the science and engineering fields, the number of these graduate students holding research assistantships, and the estimated number of full-time equivalent scientists and engineers engaged in R&D from 1974 to 1979. In 1979, of the 224 thousand U.S. full-time graduate students in science and engineering fields, 48.5 thousand held research assistantships. The latter figure is reduced by NSF to 21.0 thousand full-time equivalent scientists and engineers employed in R&D, or about 9.4 percent of the total full-time science and engineering graduate student enrollment.

The research and development effort of the other nonprofit institutions sector is fairly equally divided among basic research, applied research, and development. Although current data on the breakdown by field of employment of R&D scientists and engineers in this sector are not available, the National Science Foundation estimates that more than one-half of the R&D personnel in this sector were life scientists and engineers.[28]

As noted above, the individual R&D performer sectors in the United States are not directly comparable to those in the Soviet Union, due to wide differences in the institutional characteristics of the two systems. While higher educational institutions in the United States conduct mostly basic research, accounting for more than 60 percent of total basic research in the United States, the Academy of Sciences Institutes in the Soviet Union are the principal sector in which basic research is performed. These institutes, for which there is no analogue in the U.S. system, account for 67 to 80 percent of the basic research in the USSR. Soviet higher educational institutions, on the other hand, conduct a relatively small percentage (less than 5 percent) of the total Soviet R&D effort, and only about 10 to 13 percent of the basic research. There is some parallel, however, between Soviet branch scientific organizations and the U.S. industry sector. Soviet branch institutes primarily conduct applied R&D, and account for over 85 percent of the Soviet total. Similarly, U.S. industry performers account for about 60 percent of the total U.S. applied R&D effort.

---

[27]Discussions with NSF personnel.
[28]National Science Foundation, *National Patterns of R&D Resources, 1953–1977*, NSF 77-310, p. 16.

TABLE VII-20

U.S. GRADUATE SCIENCE AND ENGINEERING ENROLLMENT, RESEARCH ASSISTANTSHIPS,
AND AS FULL-TIME EQUIVALENT SCIENTISTS AND ENGINEERS ENGAGED IN R&D:
1974-1979

(In Thousands)

	Total Full-Time Graduate Science and Engineering Enrollment[1]	Graduate Student Research Assistantships	Graduate Student Full-Time Equivalent Scientists and Engineers in R&D[2]
1974	195.9	39.6	17.5
1975	210.8	40.1	18.6
1976	214.7	42.7	18.9
1977	218.4	43.9	20.1
1978	216.8	NA	20.6
1979	224.1	48.5	21.0[3]

[1] At doctorate-granting institutions

[2] Number of FTE graduate students receiving stipends and engaged in R&D

[3] Estimate

Sources: National Science Foundation, Academic Science: Graduate Enrollment
          and Support, Fall 1979, NSF 80-321, pp. 48 and 60; National Pattern
          of Science and Technology Resources 1980, NSF 80-308, p. 33.

CHAPTER VIII

Mobility of Scientists and Engineers

The utilization of scientists and engineers within an economy, either market or planned, is a function of how that system allocates trained persons among the respective job openings and how the system reallocates these persons through various forms of mobility. An analysis of the utilization of scientists and engineers reveals both static and dynamic dimensions. At any point in time scientists and engineers are employed in specific positions. However, over time, the number and composition of the scientists and engineers change as new entrants join the system and those already in the system seek different jobs, different regions, or perhaps even different fields or disciplines. Thus, a dynamic dimension of utilization of scientists and engineers comes under consideration.

In this chapter, the mobility of scientists and engineers within the United States and the Soviet Union will be discussed and contrasted. Several types of mobility, such as between geographical areas, between fields, and between positions, will be examined and those aspects of the economic system that facilitate or discourage mobility will be discussed.

A. Mobility in the USSR

Soviet citizens view the avenue to upward economic mobility as a function of their education, their affiliation with or compatability with the Communist Party, the general societal goals which it determines, and the ability to combine these factors to obtain positions which are ranked high in social and economic esteem—and are rewarded with both prestige and rubles. It is within this broader sense of mobility that we turn to investigate more specific forms of mobility, i.e., mobility which occurs after the labor entrant has completed his education or training and obtained his first position. The Soviet perspective of mobility is that it functions to provide a better allocation of the scientific and technical work force to achieve the national economic and scientific objectives.

By the admission of Soviet scholars, Soviet research on mobility is at the beginning stages. Although they have identified a number of key questions with respect to mobility, they only began data collection and analysis in the 1970s. Therefore, the discussion which follows will reflect the Soviet stage of development in this important area of utilization of scientists and engineers and the fact that there are extremely limited data for comparing and contrasting with the United States. The reader should be alerted to the fact that even in those cases where there are data, they generally represent a relatively small sample of the population and the conclusions which are drawn must be read with this caveat in mind.

Soviet scholars have correctly surmised that the study of mobility of personnel in science and engineering is really the study of a process—a process which has causal links and causal consequences, both societal and economic. Soviet researchers distinguish between what they consider to be essential and excess or superfluous mobility. Essential mobility is defined as that which results in growth in the number and level of qualification of scientists and engineers according to planned goals. Excess or superfluous mobility is claimed to have little impact on the development of science or individual skills, but rather simply reflects individual preferences for job change or field change and may represent movement into fields which are already adequately supplied. The former mobility is encouraged by the government, the latter is discouraged. The Soviet leadership has not as yet developed a management mechanism to guide mobility toward state-determined goals, but it has initiated attempts to analyze the mobility process so as to begin designing such a management mechanism.

Soviet analyses of mobility to date have focused on two general categories of mobility: professional and geographical.

1. Professional Mobility

Professional mobility can be defined as mobility in which the scientist or engineer has changed the subject or the methods of his research. The change may require retraining (either formal or informal), a change in specialization, in specialty or in the profession itself. Professional mobility may involve the departure of the scientist or engineer from the bounds of his or her own narrow specialty, his use of methods or concepts belonging to other disciplines or "sciences," or his investigation of new or different research topics. Thus, professional mobility includes inter- and intra-professional shifts.

Soviet researchers admit that "professional mobility" may be by intent on the part of the engineer or scientist or simply be a consequence of the absence of positions in the field or specialty in which the scientist or engineer was trained. However, it is not clear in the Soviet literature as to whether the fact that scientists and engineers are not employed in the specialty in which they were trained is a consequence of inadequate planning and/or poor government allocation or reflects personal choices by these trained professionals to locate in a region or work in a particular field other than that of their primary training. Soviet students sometime study in a field even though it is not their special interest, because higher educational institutions did not admit them in the field of their first choice. The Soviet data and analyses have not or are unable to draw a distinction between planner-induced and individual-determined professional mobility.

Soviet researchers have attempted to classify types of mobility. For instance, scientists and engineers are classified as relatively stable if they are employed in fields of their specialty training or in close approximation to that field. On the other hand, the scientist or engineer is classified as more mobile as the field in which the person is employed is further removed from the field or specialty of training.

The key characteristics or extent of professional mobility have been identified as:

- Distance from educational specialty
- Newness of the scientific field
- Changes in characteristics and types of research

a. Distance from Educational Specialty

A study involving 1,400 scientific workers of the Academy of Sciences of the USSR and those of the Union Republics generated the following data regarding the degree of correspondence between areas of employment and field of specialization in higher education:[1]

---

[1]Gvishiani et al., op. cit., p. 53.

	Share in percent
Correspond fully to the narrow specialty studied in higher educational establishments	27.7
Do not correspond to the narrow specialty studied in higher educational establishments, but correspond to the specialty group of their diploma	42.8
Work in a field of science connected with the specialty of their diploma	23.7
Correspond neither to the specialty nor the specialty group of their higher education and do not work in a related field of science	5.8

Although roughly 70 percent of the scientific workers surveyed were researching in their narrow specialty or their specialty group, still about one out of every four workers was researching outside his specific area of training.

Another Soviet study had a sample of about 2,000 scientific workers of branch scientific research institutes which specialized in work in the fields of chemistry and chemical technology. The questionnaire used in that study included questions about the educational specialty and field of present occupation of those interviewed. A single classification of specialties was provided which grouped into categories 16 specialties in the field of chemical sciences (group I), 34 in the field of chemical technology (group II), other technical sciences not divided into specialties (group III), physical sciences (group IV), and other sciences. (A very insignificant number of those interviewed were included in the latter category, which was therefore excluded from further examination.) Of the total number of those interviewed who fell in groups I to IV (1,716 individuals), a majority had the same educational and present occupational specialty. The distribution is shown in Table VIII-1.

The authors pointed out that this particular study represents a comparison of two highly heterogeneous fields.[2] Mobility between related

[2]In the USSR Nomenclature of Scientific Specialties, chemistry is an independent branch of science, while chemical technology is listed as a subspecialty under the "technical sciences" branch (which represents research engineering). In this sense they are separate and heterogeneous fields.

TABLE VIII-1

CORRELATION BETWEEN FIELD OF TRAINING AND PRESENT FIELD OF
OCCUPATION OF 1716 SCIENTIFIC WORKERS AT U.S.S.R. BRANCH SCIENTIFIC
RESEARCH INSTITUTES OF A CHEMICAL AND CHEMICAL-TECHNOLOGICAL NATURE[1]

A. Distribution of Persons Trained in Chemistry, Chemical Technology,
    Other Technical Sciences and Physical Sciences, by Present Field
    of Employment

Field of Training	Number Trained	Present Field of Employment (In Percent)			
		Chemistry	Chemical Technology	Other Technical Sciences	Physical Sciences
Chemistry	332	65.7	30.1	0.6	3.6
Chemical Technology	1113	14.4	85.0	0.4	0.2
Other Technical Sciences	169	5.9	27.8	63.9	2.4
Physical Sciences	103	53.0	10.8	1.0	35.3

B. Distribution of Persons Employed as Chemists, Chemical Technologists,
    Other Technical Scientists and Physical Scientists, by Field of
    Training

Present Field of Employment	Number Employed	Field of Training (In Percent)			
		Chemistry	Chemical Technology	Other Technical Sciences	Physical Sciences
Chemistry	442	49.3	36.2	2.3	12.2
Chemical Technology	1104	9.1	85.7	4.2	1.0
Other Technical Sciences	116	1.7	4.3	93.2	0.9
Physical Sciences	54	22.0	3.7	7.4	66.7

[1] Data compiled from selective questionnaires.

Source: S.A. Kugel' and V.S. Maslennikov, "The Mobility of Scientific
         Personnel: Supplement to the Survey Report on 'Scientific
         Personnel in the USSR'," Working Material, Moscow, p. 20 (July 1977).

specialties could be much higher than between groups of specialties (fields).

b. Mobility to New Fields of Science

With respect to the second criteria for classifying the direction of mobility—newness of the scientific field—the following types of mobility have been identified: (1) from a traditional scientific field to a new one; (2) from one traditional field to another traditional one; (3) from one new scientific field to another new one; and (4) from a new scientific field to a traditional one. Data indicate that the largest number of scholars move into new fields, but a significant number also move between the various traditional fields. (See Table VIII-2.)

The rapid development of Soviet industry and Soviet science has resulted in the opening of many new scientific and engineering fields. As new fields were developed, new positions were created, sometimes within existing institutions, sometimes at wholly new facilities. A comparison of scientific specialties for different years reveals the extent to which science has developed and become differentiated in the Soviet Union. In the 1970s, new specialties appeared in connection with research on plasma, radiochemistry, molecular biology, geology of the ocean, the development of computers, allergology and immunology, etc. As a rule, increased specialization has resulted in the formation of new, more narrowly defined occupational groups. For example, psychology has been differentiated from a teaching specialty to become an independent branch encompassing 11 specialties. The number of specialties for which degrees can be awarded increased from 72 in 1969 to 122 in 1972.

Just as new fields have been developed, among which mobility became a possibility, so too have new scientific establishments. Since 1950 the total number of scientific establishments has risen from 3,447 to 5,327 in 1975. The number of scientific research institutes and their branches and departments has more than doubled while higher educational institutions declined between 1950 and 1960 and then rose gradually to 856 in 1975—still 24 establishments below the number in 1950.

The consequence of the opening of new fields and new scientific establishments must have greatly enhanced the potential mobility of scientists toward new fields and new establishments. Undoubtedly, many of the new scientific and engineering fields and specialties were filled by new entrants. There must have also been, however, a significant number of experienced scientists who were attracted to these

TABLE VIII-2

PERCENTAGE DISTRIBUTION OF ANSWERS OF USSR SCIENTIFIC WORKERS
TO THE QUESTION: "IS THE LAST CHANGE OF YOUR SPECIALTY
CONNECTED WITH A TRANSFER TO A NEW SCIENTIFIC FIELD
OR DOES IT IN ESSENCE REPRESENT A TRANSFER TO ONE
OF THE TRADITIONAL SCIENTIFIC FIELDS?"[1]

From a traditional scientific field to a new one	34
From a new scientific field to another new one	21
From a new scientific field to a traditional one	2
From one traditional scientific field to another traditional one	43

---

[1] Data compiled from responses to questionnaires completed by
scientific workers at a group of branch scientific research
institutes in Leningrad.

Source: S.A. Kugel' and V.S. Maslennikov, "The Mobility of Scientific
Personnel: Supplement to the Survey Report on "Scientific
Personnel in the USSR'," Working Material, Moscow, p. 18
(July 1977).

opportunities and were either retrained or self-trained as prerequisites to movement from an old field to the new developing fields.

Soviet research on who changes fields reveals that older groups tend to make up better than half of those persons changing specialties. This is explained by the fact that new fields of research are usually developed by scientists who are leaders in their field and thus more mature and experienced.

c. Changes in Characteristics and Types of Research

A special type of mobility to be considered is that between theoretical and experimental work. Sociological research indicates that in branch scientific research institutes and, to a certain extent, in academic establishments, there has been a trend toward an increase in the amount of experimental work. At the same time, the personal orientation of a significant number of scientific workers is primarily or exclusively directed toward theoretical work. The reasons for this inconsistency and the means to overcome it deserve special attention.

With respect to changes in types of scientific activity—mobility between fundamental and applied research—mobility generally occurs between related types of research (e.g., applied research to field work). On the average, scholars make this kind of change more than twice, but the tendency is to move closer to production work. (See Table VIII-3.)

One should be cautious in generalizing from the sample identified in Table VIII-3. However, the conclusions drawn by those conducting the research are consistent with the general trends in the Soviet economy and in science and technology specifically. The change in the character and types of research reflects the design of the leadership to shift their scientific and engineering personnel to research problems associated with overcoming bottlenecks in the economy.

2. Geographical Mobility

Geographical mobility may result from government action to allocate scientific and engineering manpower within and among the republics or it may reflect individual preferences to relocate to further career objectives or to work in more desirable geographical areas. Unfortunately, Soviet research on this topic has not sufficiently isolated the various causal factors which result in geographical mobility to distinguish between these broad categories.

One set of data shows that professional mobility sometimes involves

TABLE VIII-3

PERCENTAGE DISTRIBUTION OF ANSWERS OF USSR SCIENTIFIC WORKERS TO THE
QUESTION: "IF THE TYPE OF YOUR RESEARCH CHANGED, THEN IN WHICH
DIRECTION?"[1]

Direction of Mobility

From applied to fundamental research	9.2
From developing new technology and techniques to fundamental research	6.6
From developing new techniques and construction design work to applied research	7.9
From industrial start-up to applied research	15.8
From industrial start-up to construction design work	3.9
From fundamental to applied research	11.8
From fundamental research to development	3.9
From fundamental research to construction design work	1.3
From applied research to development	14.5
From applied research to construction design work	5.3
From applied research to industrial start-up	14.5
From construction design work to industrial start-up	2.6
From research activity (exclusively) to scientific-pedagogical work	1.3
From scientific-pedagogical work to research activity	1.3

---

[1] Data compiled from selective questionnaires completed by employees
    of a branch scientific-research institute.

Source: S.A. Kugel' and V.S. Maslennikov, "The Mobility of Scientific
        Personnel: Supplement to the Survey Report on 'Scientific
        Personnel in the USSR'," Working Material, Moscow, p. 19 (July 1977).

a change in place of employment or a transfer to another district of the country. While this type of mobility is sometimes related to a scientist's or engineer's change in specialty or specialty group, migration usually does not occur when a scientific organization in the new specialty is created. The reason for this is that there are many cases in which the scholar is able to conduct research using the new technique in an existing laboratory. The results of questionnaires on factors related to changes in place of employment indicate that dissatisfaction with scientific specialty is a relatively rare reason given for geographical mobility: only 4 to 6 percent of those interviewed in branch institutes listed this as a factor (the percentage was more significant for the most highly qualified scientific workers). However, in cases in which a scholar changed his specialty while at one place of employment and thereafter moved to a new place of employment where he continued to work in his new specialty, he did not have the option of listing dissatisfaction with specialty as a reason for this displacement, although it may very well have been a reason. Thus, dissatisfaction with specialty may be a more significant factor in changes in place of employment than the results of the questionnaire would indicate.

Expansion of new scientific institutes and thus new job opportunities seems to be primarily in the non-European sections of the Soviet Union. One would expect that some of these positions will be filled with new entrants into these professions or fields, but the core cadre, particularly the administrative and laboratory heads, of any new institute must come from the pool of experienced personnel, and, therefore, there must be significant numbers of trained persons moving into these geographical areas because of the opportunities there which are not available in the established institutes in European U.S.S.R. On the other hand, one might hypothesize that these experienced scientists would not elect to move east for reasons of location. Unfortunately, there are few data to substantiate either hypothesis.

3. Economic Incentives and Mobility

A review of the various sources of research on mobility in the Soviet Union has shown that Soviet scholars have almost totally ignored the importance of economic incentives as a causal factor in mobility or even as a facilitator in achieving desired changes in the distribution of Soviet scientists and engineers among the various fields and specialties, at the different types of scientific institutions, and among the geographical regions. In two separate conferences on training and utilization of scientists and engineers sponsored by the Science Policy Working Group of the U.S.-U.S.S.R. Agreement for Cooperation in the Fields of Science and Technology, this issue was raised for discussion.[3] The Soviet attendees' response to a question regarding the role of economic incentives, particularly salaries, as a means for encouraging scientists and engineers to change fields was that salaries were not used to encourage mobility between fields or within fields of science. There are some salary differentials among regions which are built into the basic wage structure. The structure, however, is not altered for specific individuals to entice them to accept employment at a particular scientific or industrial institution.

The Soviet scholars contended, first, that scientific or engineering positions, either existing or newly created, are filled through the functioning of the plan. For instance, they contended that a new area of science and technology is never developed *ex nihilo*. Everything new arises out of something that previously existed. Therefore, when a need arises for specialists in a new field of science or technology, the financing and resources for training individuals in the new field are increased. Such considerations, therefore, should be accounted for in the plans. The plan would define the institutions at which the goals can best be achieved, and establish mechanisms for training the required number of specialists. This response still leaves unanswered how, through mobility, either individual-initiated or state-induced, scientists and engineers are reallocated as state goals and objectives change or as there are scientific and technological advances generating new fields. While new scientists and engineers are being trained, presumably according to plan to meet these new needs, scientists and engineers already trained must be reallocated to fill these needs. But how does the system attract or redirect them?

According to the Soviet scholars at the Moscow conference on training and utilization of scientists and engineers, the scientists working at established institutions in the prime locations (Moscow or Leningrad and other Western cities) typically join the institution's staff and stay in its employ in hopes of being promoted to the higher positions. Since the turnover is so low and the director, once appointed, remains in that position until retirement, the upward movement has been tradi-

[3]See Catherine P. Ailes, "Report on the Conference of the U.S.-U.S.S.R. Joint Subgroup on the Training and Utilization of Scientific, Engineering and Technical Personnel: March 8, 9 and 10, 1978," SRI International (May 1978); and Francis W. Rushing, "Report on the Moscow Conference of the U.S.-U.S.S.R. Joint Subgroup on the Training and Utilization of Scientific, Engineering and Technical Personnel: June 26, 27 and 28, 1978," SRI International (August 1978).

tionally slow. A director, on the other hand, has few opportunities to bring in new scientists or engineers because so few positions are vacated. Thus, to attract new talent, particularly experienced scientists, the director attempts to justify new positions which he can fill with more senior persons. During *informal* conversations it was admitted that in fact economic and quasi-economic incentives were used to attract new and proven personnel. The most important incentive was housing. If the scientific institution director either has housing associated with his institution, or through official or unofficial allocation has access to a prime apartment, he has an important attraction for a potential employee. Discussions with other directors also revealed that for newly graduated scientists and engineers, apartment space is a key inducement. Thus, an institute director may or may not be able to manipulate salaries to attract workers, but he must, to be competitive, be able to manipulate the fringe benefits, particularly apartment space. Some scientific institutions which are prestigious within the Soviet scientific community may be able to attract senior scientists without special inducements. There are cases of scientists working in Moscow with their families in the Ukraine or Leningrad, awaiting an apartment for their transfer.

B. Mobility in the United States

Mobility of scientists and engineers in the U.S. market economy is traditionally analyzed within the context of the theory of human capital. The analysis holds that individuals decide to move between firms, occupations, fields, institutions, or geographical regions after evaluating the expected returns and goals (both monetary and nonmonetary), and the risk associated with the move. The greater the expected returns, the greater the incentive for mobility. Thus, if expected salaries and employment levels in a different field are relatively large, an individual has a greater incentive to transfer fields.

Mobility may be required of an individual by unexpected changes in technology or in the demand for goods and services. The introduction of new products or shifts in demand may invalidate employment and salary expectations. Unanticipated changes in market demand are often responded to by retraining and mobility. Even if technology were constant, mobility is desirable because it allows individuals and firms to correct past mistakes on the basis of current knowledge. Consequently, mobility aids efficient resource allocation.

There are costs to mobility, but they are generally short-run costs. For instance, there are costs to the firm in obtaining information about trained persons, costs to the individual in obtaining information about available jobs, and costs of moving the individuals and their families. More importantly, there is the loss in output. Because most jobs require some familiarization with the process or organization of the firm, an employee will not function at his maximum level of efficiency immediately upon assuming his new position. There may also be a loss in income if there is a time gap between jobs.

1. Field and Occupational Mobility

Postcensal data are used to analyze the direction and extent of field and occupational mobility of scientists and engineers.[4] Because a number of criteria were used to define scientists and engineers for the survey, it is expected that the percentages underestimate those whose occupations differed from the field of specialty pursued in their formal schooling.

a. Engineers

Table VIII-4[5] shows the percentage of engineers who remained within their own field (retention rate), and the percentage that were working in occupations other than their own field. As shown in this table, the retention rate for engineers employed as engineers in 1972 was 80.1 percent. Another 15.2 percent became administrators and managers, and 0.3 percent became teachers. Thus, only 4.4 percent of engineers in 1972 were known to go into science and other fields. Those with the greatest mobility to fields outside of engineering were chemical engineers (6.1 percent), and those with the lowest mobility were agricultural engineers (0.9 percent).

For engineers working as engineers, retention rates varied between 70.4 percent for chemical engineers and 80.0 percent for electrical and electronic engineers. (The inclusion of the occupation "environmental and sanitary" into "civil and architectural" increases the retention rate of civil, environmental, and sanitary engineers to 81.3 percent.) Although 13.1 percent of chemical engineers went into administration or teaching, 10.4 percent transferred to other fields of engineering, and 6.1 percent transferred to fields outside of engineering.

Engineers moving from one engineering field to another engineering

---

[4]See Chapter VI, Section B, for a discussion of the methodology used in the Postcensal Survey.
[5]It should be noted that the data in Tables VIII-4 and VIII-5 are not standardized for age or years in professional employment. In most cases, the difference would be fairly small, but for rapidly changing fields and new fields, it might be significant.

TABLE VIII-4

DISTRIBUTION OF U.S. ENGINEER POPULATION, BY OCCUPATION AND FIELD: 1972
(In Percent)

Occupation	Total	Aeronautical and Astronautical	Agricultural	Chemical	Civil, Environmental and Sanitary	Electrical and Electronic	Industrial	Mechanical	Metallurgical & Materials	Mining and Petroleum	Nuclear
Total employed	100.0	100.0	100.0	100.0	100.0	100.0	100.0	100.0	100.0	100.0	100.0
Engineers, total	80.1	88.9	95.8	80.8	88.6	87.3	82.1	86.6	86.1	80.6	89.6
Aeronautical, astronautical	4.0	77.2	0	0.1	0.2	0.9	0.2	1.5	0.4	0	c
Agricultural	0.5	0.2	79.3	0.4	0.1	0.1	0	0.2	0	0	c
Chemical	4.1	0	0.6	70.4	0.1	0.1	0.1	0.4	1.0	0.9	c
Civil and architectural	11.4	0.1	0.7	0	74.0	0.1	0.1	0.5	0.1	0.3	c
Electrical and electronic	19.4	1.4	0	0.6	0.3	80.0	1.7	0.7	0.4	c	c
Industrial	4.7	0.1	2.8	0.5	0.5	0.5	73.5	0.8	0	c	0.6
Mechanical	15.9	4.7	3.9	0.8	0.8	0.2	1.9	73.6	1.6	1.4	3.3
Metallurgical and materials	2.2	0.4	0.6	0.7	0	1	0.1	0.5	78.5	0.7	c
Mining and petroleum	1.4	0	0	0.4	0.1	[1]	0	0.4	0.1	74.8	0
Nuclear	0.8	0	0	0.6	0.1	0.1	0.1	0.5	0	0.5	79.4
Environmental and sanitary	1.5	0.1	0	0.6	7.3	5.1	0	0.6	0	2.0	2.4
Other	14.1	4.8	7.9	1.9	5.1	1.7	4.1	7.0	3.5	0.8	4.0
Computer specialists	0.8	0.4	0.9	0.8	0.3	0.2	0.5	0.2	0	0.5	c
Mathematical specialists	0.2	0.4	0	0.3	0.1	0.3	0.6	0.3	0.3	0.4	c
Natural scientists	0.1	0	0	2.0	0	0	0.2	0.1	c	c	c
Social scientists		0	0	0	0	0	0.2	0.1	c	c	c
Technicians, total	1.5	0.8	0	0.6	1.7	2.0	1.4	1.8	0.4	0.6	3.7
Teachers	0.3	0.3	0	0.2	0.1	0.5	0.4	0.2	0	c	c
Administrators and managers	15.2	7.1	3.3	12.9	8.0	6.4	12.2	8.5	12.1	15.1	6.2
Administrators and managers, scientific research	3.7	2.9	0	3.1	0.5	1.5	1.8	1.9	4.0	1.1	0.4
Administrators and managers, production/operations	5.4	1.3	0	5.1	2.4	2.2	5.9	3.0	3.4	10.6	5.8
Other administrators and managers	4.5	2.1	0.7	2.6	3.6	1.8	3.4	2.8	3.7	2.9	c
Self-employed proprietors	1.4	0.7	2.6	2.0	1.5	0.9	1.0	0.9	1.1	0.5	c
Other occupations	1.6	2.2	0	2.1	1.2	1.7	2.4	2.1	1.0	1.9	0.4

[1] Less than .05 percent.

Note: Detail may not add to 100.0 percent because of rounding.

Source: Calculated on the basis of data contained in "The 1972 Scientist and Engineer Population Redefined" (Engineers, by Field), National Science Foundation, NSF 76-306, p. 1 (1976).

field ranged from a low of 5.8 percent for mining and petroleum engineers to a high of 16.5 percent for agricultural engineers. Transfers among engineering fields were made most frequently *into* mechanical and electrical and electronic engineering and most frequently *from* aeronautical, agricultural, civil, and mechanical engineering.

b. Scientists

Table VIII-5[6] shows that the retention rates of scientists employed as scientists varied between 66.9 percent (chemists) and 87.5 percent (medicine). Operations research analysts had the highest mobility into administration and management (23.3 percent), which is a logical extension of the field. Other science fields with relatively greater movement from science into administration and management were economics (14.2 percent), chemistry (13.8 percent), agricultural science (11.2 percent), computer science (11.0 percent), and statistics (11.0 percent). Chemists had the largest percentage employed in management of scientific research. Persons trained in medical and computer sciences had the highest retention rates (87.5 percent and 87.2 percent, respectively), while those trained in chemistry and operations research were relatively more mobile with retention rates of 66.9 percent and 69.0 percent, respectively. With the exception of scientists employed as teachers or administrators, those with specialties in biology, physics, and astronomy have the highest mobility rates into other fields (15.9 percent and 14.6, respectively).

c. Field Mobility of Ph.D.s

Detailed information on field mobility of Ph.D.s was collected in a 1973 survey of a representative sample of doctoral scientists and engineers conducted by the National Academy of Sciences (NAS).[7]

Various aspects of field mobility had been dealt with in previous NAS reports, and data show that the number of individuals switching from their doctoral specialties to different fields of employment had been on the increase since 1960. By 1973, one out of every six employed doctoral scientists and engineers had changed his field of employment. Those who changed their fields were more frequently em-

[6]See footnote 5.
[7]*Field Mobility of Doctoral Scientists and Engineers*, Commission on Human Resources, National Research Council, National Academy of Sciences (December 1975).

TABLE VIII-5

U.S. DISTRIBUTION OF SCIENTIST AND ENGINEER POPULATION, BY OCCUPATION AND FIELD: 1972

(In Percent)

Occupation	Engineering	Mathematics	Statistics	Computer Science	Operations Research	Agricultural Science	Biology	Chemistry	Medical Science	Physics/Astronomy	Psychology	Economics	Sociology/Anthropology
Total employed	100.0	100.0	100.0	100.0	100.0	100.0	100.0	100.0	100.0	100.0	100.0	100.0	100.0
Engineering	80.1	3.0	1.5	0.4	1.6	2.4	0.9	1.6	0.2	6.9	0.1	0.5	0.8
Mathematics	[1]	79.7	1.0	0.1	0.5		0.1	0.1		1.0		0.7	
Statistics	[1]	0.8	79.7		0.3	0.2	[1]				0.2	0.6	0.3
Computer science	0.6	4.3	[1]	87.2	3.2				0.2	1.0	0.1	0.5	0.6
Operations research	0.1	1.2	[1]	0.1	69.0	[1]	0.2	0.2		0.5		0.2	0.5
Natural science	0.2	0.5	1.4	0.2	0.5	80.1	84.0	79.3	89.9	78.5	0.2	0.3	0.3
Agricultural						78.2	1.5						
Biology						0.9	76.0	0.5	0.7	0.5			
Chemistry	0.1					0.1	0.9	66.9	0.7	0.5			
Medical						[1]	2.1	0.8	87.5		0.5	0.3	0.2
Physics		0.2					0.1	0.2		74.8			
Social science			0.7		0.5	0.1		0.1	0.2		85.0	75.1	78.9
Psychology			0.5								83.5		
Economics			0.3			0.1					0.2	74.5	
Sociology/anthropology			1.7										76.8
Teacher	0.3	2.5	[1]	[1]	0.7	0.6	2.9	1.4	1.1	2.5	5.0	5.1	4.8
Technician	1.5	0.2		0.2	[1]	1.9	4.7	1.0		0.5	0.2		
Administration and management, total	15.2	4.8	11.0	11.0	23.3	11.2	5.2	13.8	3.8	8.1	7.3	14.2	9.1
Scientific research	3.7	1.1	0.7	2.8	6.3	1.9	1.9	6.8	1.7	6.1	1.2	3.7	2.4
Production	5.4	2.1	2.0	3.0	2.5	4.3	1.1	3.0	0.2	0.8	0.7	1.9	
Other	4.5	1.4	7.4	4.8	14.2	4.1	1.1	2.8	1.0	0.8	4.8	7.0	4.9
Self-employed	1.4	[1]	1.0	0.3	0.3	0.2		0.7	0.2		0.2	0.8	
Other occupations	1.7	2.7	2.7	0.5	0.9	3.2	1.3	2.1		1.6	0.9	3.3	3.9
Not available	0.1	0.3	0.3	0.1		0.3	[1]	0.1			0.3		1.1

[1] Less than 0.1 percent.

Note: Total may not sum to 100.0 because of rounding and omissions.

Source: Calculated on the basis of data contained in "The 1972 Scientist and Engineer Population Redefined," National Science Foundation, NSF 75-327, p. 46 (September 1975).

ployed in managerial and administrative positions and, not surprisingly, generally earned more than those who remained within their own field.

This study also indicated that, as the number of years passed since receiving the Ph.D., there was an increase in the percentage changing fields. For example, the number of persons entering and leaving engineering fields was small for recent graduates but was significant for those who had received their Ph.D. 20 years or more before.

Data show that the field with the greatest proportional influx of scientists from other fields was earth sciences (56 percent), followed by mathematics (30 percent) and psychology (21 percent). It was also found that more than one-quarter of the Ph.D. chemists switched out of their field of study, and that substantial portions of physicists and social scientists were employed in fields other than those for which they received their doctorates (29 percent and 18 percent, respectively).

2. Mobility among Firms

Postcensal data give the length of tenure of scientists and engineers in their 1972 jobs by field and by age. Data in Table VIII-6 show that the average length of job tenure increases as the age of the scientist or engineer increases. For example, 17 percent of engineers under 30 had been on the job less than one year. This was true of only 11 percent of the engineers between the ages of 30 and 39, 8 percent of those between 40 and 49, and 6 percent for those 50 years of age or older. This result is not surprising because the costs of mobility increase as the lifetime rewards to mobility decrease with age. Additionally, mobility is a form of job search and the older the employee, the more likely it is that he has successfully completed his job search.

The number of scientists, even in the oldest age group—50 years of age and older—who spent 10 years or more on the same job never exceeded 57 percent. For physical and life scientists, 56.9 percent and 56.5 percent, respectively, had been on the job 10 years or more. In all other fields this percentage was lower.

The foregoing data appear to confirm the concept of vocational development as a continuous process of adaption of individual workers to the labor market. Vocational development theory asserts that "basically the individual does not choose an occupation, but rather makes a series of occupational and occupationally-related choices at different life stages."[8]

---

[8]Joseph Zaccaria, *Theories of Occupational Choice and Vocational Development*, Boston: Houghton, Mifflin Co. (1970).

THE SCIENCE RACE

TABLE VIII-6

LENGTH OF TENURE IN 1972 JOB OF U.S. EMPLOYED SCIENTIST AND ENGINEER POPULATIONS,
BY BROAD FIELD GROUPS AND AGE

Field and Age	Total Number	Less than 1 year	1-2 years	3-4 years	5-9 years	10 years or more	Not Reported
Engineers: Total[1]	799,935	77,358	176,662	127,246	162,910	198,301	57,459
Under 30[1]	122,691	21,228	43,486	33,473	17,623	683	6,199
30-39 years[1]	253,551	26,551	66,764	47,360	62,366	30,893	19,617
40-49 years[1]	247,569	18,896	42,854	32,505	54,110	81,377	17,828
50 years & older[1]	176,124	10,683	23,559	13,908	28,811	85,348	13,815
Mathematical specialists: Total[1]	29,460	3,099	7,333	5,877	6,622	5,327	1,203
Under 30[1]	7,212	1,422	3,027	1,981	656	0	125
30-39 years[1]	10,948	1,193	2,987	2,480	3,016	658	614
40-49 years[1]	6,641	293	860	961	2,027	2,275	225
50 years & older[1]	4,659	191	459	455	922	2,394	239
Computer specialists: Total[1]	97,698	16,256	31,992	19,737	16,782	8,132	4,800
Under 30[1]	28,362	5,515	12,170	6,516	3,282	74	805
30-39 years[1]	45,385	7,514	14,761	8,752	9,080	2,590	2,689
40-49 years[1]	19,337	2,756	4,412	3,834	3,606	3,698	1,031
50 years & older[1]	4,615	471	649	635	814	1,770	275
Operations research analysts: Total[1]	11,305	1,723	3,677	2,135	2,002	933	835
Under 30[1]	3,143	505	1,603	532	200	0	303
30-39 years[1]	4,532	604	1,077	893	1,241	339	377
40-49 years[1]	2,523	475	779	633	386	226	24
50 years & older[1]	1,107	139	218	77	175	368	130
Life scientists: Total[1]	72,757	9,699	16,054	12,584	14,486	17,306	2,628
Under 30[1]	12,898	3,992	5,319	2,361	809	159	258
30-39 years[1]	26,329	3,336	6,792	6,178	6,451	2,462	1,111
40-49 years[1]	18,454	1,450	2,597	2,737	4,799	6,172	700
50 years & older[1]	15,075	921	1,346	1,309	2,427	8,513	559
Physical scientists: Total[1]	170,098	17,794	35,484	28,978	36,817	42,879	8,146
Under 30[1]	29,470	6,645	11,206	6,633	4,163	0	823
30-39 years[1]	59,807	6,458	14,989	13,210	15,928	6,182	3,041
40-49 years[1]	45,022	3,329	6,515	5,733	10,502	16,319	2,625
50 years & older[1]	35,798	1,363	2,773	3,402	6,224	20,379	1,657
Psychologists: Total[1]	34,653	5,900	9,073	6,477	6,316	4,156	2,731
Under 30[1]	8,267	2,964	3,124	1,659	119	0	402
30-39 years[1]	11,499	1,796	3,658	2,652	2,127	363	904
40-49 years[1]	9,125	884	1,606	1,374	2,522	1,948	791
50 years & older[1]	5,761	256	686	792	1,548	1,845	634
Social scientists: Total[1]	55,957	8,560	16,563	10,822	9,967	7,296	2,748
Under 30[1]	12,660	3,606	5,736	2,004	574	0	740
30-39 years[1]	18,319	3,097	5,669	4,856	3,247	577	873
40-49 years[1]	13,514	1,351	2,882	2,648	3,765	2,253	615
50 years & older[1]	11,464	507	2,276	1,313	2,382	4,465	521

---

[1] Includes persons not reporting occupation

Source: "The 1972 Scientist and Engineer Population Redefined," National Science Foundation, NSF 75-327, pp. 98-100 (September 1975).

3. Geographical Mobility

The postcensal data give the distribution of engineers and scientists by geographical region of employment in 1972 but do not provide regions of birth, schooling, or previous employment. Other studies have found that geographic mobility depends significantly on whether the labor market for a particular skill is national or regional.[9] Another study has found that movements of scientists and engineers have tended to follow patterns of the general population.[10]

A study by the National Academy of Sciences has traced the mobility of a number of holders of the Ph.D.[11] The study is based primarily on data from the Doctorate Records File based on the periodic Survey of Earned Doctorates, which includes essentially all doctorates awarded in the United States and not merely those in the scientific and engineering fields. For tracing the mobility of Ph.D.s after having obtained the degree, data from the Doctorate Survey were supplemented with data from the National Register of Scientific and Technical Personnel maintained by the National Science Foundation.

Table VIII-7 presents, by geographical region, patterns observed in the movement from place of baccalaureate to place of Ph.D. in the period 1960 to 1967. As this table and the listing below indicate, approximately half of the students moved from one of the nine geographic regions of the United States to another.

Region	Percent moving to another region
New England	58
Mid-Atlantic	42
East North Central	50
West North Central	47
South Atlantic	59
East South Central	49
West South Central	38
Mountain	63
Pacific	51

[9] John K. Folger, Helen S. Astin, and Alan E. Bayer, *Human Resources and Higher Education,* New York: Russell Sage Foundation, pp. 217-253 (1970); Dael Wolfle, *The Use of Talent,* Princeton: Princeton University Press, p. 159 (1971).

[10] Shapero, Howell and Tombaugh, "The Structure and Dynamics of the Defense R&D Industry," Stanford Research Institute (1965).

[11] *Mobility of Ph.D.s: Before and After the Doctorate with Associated Economic and Educational Characteristics of States;* Career Patterns Report Number Three, Research Division of the Office of Scientific Personnel, National Academy of Sciences (1971).

TABLE VIII-7

MIGRATION OF U.S. PH.D.S: 1960-1967

Region of Baccalaureate	Region of Ph.D.									
	New England	Mid Atlantic	East North Central	West North Central	South Atlantic	East South Central	West South Central	Mountain	Pacific	Total
New England	4,033	2,105	1,463	299	663	66	131	141	937	9,838
Mid Atlantic	2,285	11,345	3,030	628	1,640	133	281	306	1,255	20,903
East North Central	1,229	1,932	11,653	1,358	1,033	186	399	591	1,578	19,959
West North Central	425	692	2,169	4,506	490	112	499	675	886	10,454
South Atlantic	513	1,199	1,227	244	3,910	405	316	137	379	8,330
East South Central	162	289	754	181	716	1,283	342	68	144	3,939
West South Central	252	350	903	468	434	256	3,815	313	415	7,206
Mountain	197	353	771	361	181	35	190	1,573	1,002	4,663
Pacific	585	1,151	1,173	391	368	50	183	450	6,353	10,704
TOTAL UNITED STATES	9,681	19,416	23,143	8,436	9,435	2,526	6,156	4,254	12,949	95,996

Source: Mobility of Ph.D.s: Before and After the Doctorate with Associated Economic and Educational Characteristics of States; Career Patterns Report Number Three, Washington, D.C.: Research Division of the Office of Scientific Personnel, National Academy of Sciences, p. 120 (1971).

More detailed data are presented by state regarding the movement of Ph.D.s from their high school location to their baccalaureate, baccalaureate to Ph.D., and Ph.D. to place of first employment. A factor analysis of the observed migration patterns showed dependence on a measure of economic prosperity, the quality of elementary education, and the quality of higher education in the state.[12]

C. U.S.-Soviet Comparison

A search of both Soviet and U.S. literature on mobility has revealed only one study of note in comparing the mobility of scientists within the two countries.[13] This article, by Linda L. Lubrano and John K. Berg, reflects some of the findings of only a part of a broader study of natural scientists in these two economies which these authors are undertaking.

The research focuses on natural scientists in the fields of physics, chemistry, and biology who were in the Soviet Academy of Sciences and the U.S. National Academy of Sciences for the years 1920-1970. The biographical data on these scientists were drawn from a variety of sources published within the respective countries. The sample size of natural scientists was 645 for the U.S. and 397 for the USSR. The sample for 1970 represented 0.19 percent of all natural scientists in the U.S. and 0.15 percent in the USSR; they represented, however, almost the entire population of natural scientists in the respective academies.

The study found that scientists of the USSR and U.S. national academies of sciences were very similar with respect to age and sex distribution. The USSR has 58.5 percent of its membership between 50 and 70 years of age, with 0.9 percent female for the year 1970. In the United States, the percentages were 61.5 and 1.1, respectively, for the same year. The members of the academies tended to be trained in the most prestigious schools, with Moscow State University and Leningrad State University dominating in the Soviet Union. The dominant fields of specialization in the Soviet Academy were physics (43.7 percent) and related technical fields. Biology (30 percent) was the largest single field in the U.S. Academy, but physics, chemistry, and biochemistry also had significant representation.

[12]Ibid.
[13]Linda L. Lubrano and John K. Berg, "Scientists in the U.S.A. and U.S.S.R.," *Survey*, Volume 23, No. 1 pp. 161-193 (Winter 1977-78).

Turning to the mobility patterns within the two countries, one pattern is the mobility between an individual's place of training and place of employment. Data presented in the study support the conclusion that there are close institutional ties in the United States as well as in the Soviet Union. For instance, the researchers at the Soviet and republic academies are drawn heavily from persons trained at universities, while research scientists at industrial enterprises come from polytechnical institutes as well as universities. In the U.S. case, there was a high incidence of "inbreeding," particularly between educators at polytechnical institutes and employment at polytechnical institutes. This general pattern holds for teachers and administrators as well as researchers in both countries. Generally, those teaching or administering in polytechnical institutes or universities were drawn from teachers and administrators from similar institutions, but usually the most prestigious.

One question of institutional mobility is the extent to which scientists move from one type of employment institution to another. The percentage of Soviet Academy scientists who had worked in only one type of institution throughout their careers was significantly larger than the percentage for the U.S. sample. If the data for earlier years are analyzed, it appears that the trend of the Soviet scientists was toward less mobility while that of the U.S. scientists was toward greater mobility.

Regional mobility is particularly important in the Soviet case, as the regime has been attempting to reduce the concentration of science in Moscow and Leningrad. While there are also concentrations of scientific activity within the United States, American scientists are generally less concentrated than their Soviet counterparts. For instance, a study of the relative degree of geographic concentration of scientists in the United States, using the Scientific Citation Index by first author's name, showed that in 1967 New York City was the address of only 5 percent of the American authors cited. The comparable percentage was 50 percent for Moscow; 33 percent for Tokyo, 26 percent for Paris, and 21 percent for London.[14]

Evidence provided by Soviet sources confirms the concentration of Soviet scientists in a few urban centers in the USSR. One Western analyst has cited a set of Soviet data showing that 34 percent of all Doctors of Science and 26 percent of all Candidates of Science lived

[14]Derek J. de Solla Price, "Measuring the Size of Science," *Proceedings of the Israel Academy of Sciences and Humanities*, Vol 9, No. 6, 1969.

in Moscow.[15] In addition, the city housed 45 percent of all scientists with the title of professor and 72 percent of the full members of the Soviet Academy of Sciences. About one-fourth of all of the scientific institutions of the USSR were located in the three cities of Moscow, Leningrad, and Kiev, but 40 percent of Soviet R&D was being performed in those cities.[16]

Soviet scientists tended to be employed in the same region where they were educated, as did the American scientists. However, there were more cases of movement outside the region where the American scientists were educated than in the case of the Soviet scientists. The important fact to note is that the American scientists generally only moved to adjacent regions.

Although there are relatively few published studies on mobility in the Soviet Union, the information pieced together in this chapter permits some generalizations concerning the differences in mobility of scientists and engineers in the United States and the Soviet Union.

There are great similarities in the two economies as to the most important forms of mobility—geographical, field, specialty. There is also a consensus that mobility of all kinds has the potential for more efficient use of highly trained and skilled manpower. The Soviets, however, differentiate mobility between "desirable" and "undesirable" mobility. The criterion for classification is whether or not the mobility facilitates achieving planned goals. In the U.S. system, voluntary mobility undertaken by an individual is assumed to be desirable, at least from the individual's perspective, since he believes he will be better off in making the move. In cases where the person is terminated for poor performance or for budgetary reasons, it would clearly be undesirable from the individual's perspective although efficiency could be enhanced for the institution.

Specialists in the scientific and engineering professions in the Soviet Union seem to be more resistant to mobility than their U.S. counterparts. For instance, the tradition of long tenures for the top positions and the strong attractions to live in particular locales such as Moscow and Leningrad and work in scientific and engineering research institutes tend to discourage mobility. Shifting of concentration of engineering and scientific activities appears to occur through the redirected flow of new entrants into these fields and locations by Soviet planners. The

[15]Paul M. Cocks, *Science Policy: USA/USSR Volume II, Science Policy in the Soviet Union,* U.S. Government Printing Office, Washington, D.C. 1980, p. 50, citing V. I. Duzhenkov, "Problemy territorialnoy organizatsii nauchnoy deyatelnosti," in *Problemy deyatelnosti uchenogo i nauchnykh kollektivov* (Moscow-Leningrad, 1977), VI, pp. 48-49.
[16]Ibid.

market tends to provide the incentives and disincentives in the United States.

Mobility in the United States is primarily guided by economic considerations—differentation of salary, more rapid advancement, fringe benefits, etc. Salaries and fringe benefits for Soviet scientists and engineers tend to be uniform. While there is evidence, although principally anecdotal in nature, that there are economic incentives functioning although often outside of official sanction, their influence appears to be less than in the United States. While some generalizations can be made concerning the mobility of scientists and engineers in the USSR, it is not possible to quantify these general impressions or even to estimate the magnitude of persons involved in the mobility process. Finally, the authors remain puzzled over how the Soviet system adapts to the ever growing need imposed by advancing technology to establish new interdisciplinary research and production teams. There does not seem to be a process similar to the rapid interfirm mobility so characteristic of U.S. advanced technology industries. Perhaps therein lies one of the professional manpower problems which is a function of Soviet institutional constraints. Unfortunately, we can only hypothesize at this time.

CHAPTER IX

Planning and Forecasting the Demand and Supply of Scientists and Engineers

Both U.S. and Soviet economists and government officials have an interest in forecasting the future supply of and demand for scientists and engineers. In the United States, economists search not only for the estimate but also for an explanation of changes in the number and composition of scientists and engineers. Government officials are concerned that there be an adequate supply of these key labor inputs. Forecasts of surpluses or deficiencies prompt a decision either to rely on the market to make the appropriate adjustments or to initiate a government policy aimed at altering the incentives or changing other market variables to enhance or diminish the supply of scientists and engineers.

In the Soviet Union, the forecast of the supply of and demand for scientists and engineers is an integral part of short- and long-term plans for the economy. As described in Chapter I, Soviet planners design and publish an annual plan, a five-year plan, and prospective plans of different durations (generally 10 to 15 or 20 years). In the design of each of these plans, manpower needs (demand) and availability (supply) must be estimated. Any discrepancies between supply and demand must be overcome by training and redirection of manpower flows. The Soviet system is heavily reliant on the effectiveness of the forecast and projections because such estimates are fed back into the planning chain. Any inadequacies in forecasting and projection procedures can have direct, and sometimes immediate, impact on the performance of the Soviet economy.

This chapter will review the general methodologies used in the two economies in forecasting or projecting the future supply of scientists and engineers. Some projections of the supply of scientists and en-

gineers in the two countries are presented to give the reader some
indication of future comparative growth rates.

A. *Methods for Forecasting Scientific and Engineering Personnel in the USSR*

The effective microplanning for the training and utilization of scientists
and engineers in the Soviet economy is predicated on a reasonably
accurate forecast which can be integrated into the plan for the economy
as a whole. To make these estimates, variables which are critical for
forecasting future needs, demands, and supplies of scientists and engineers must be identified. In this section, the methods used by Soviet
planners in making forecasts will be discussed.

There are currently several different methodologies being utilized
or developed in the Soviet Union. The most frequently used are the
nomenclature and the saturation methods. A second group of methodologies relies on statistical and mathematical techniques. The least
complex of these methodologies uses extrapolation methods while the
more sophisticated models now being developed are identified in the
Soviet literature as supply models, demand models, and balance
models.

1. Nomenclature Method

The "nomenclature of function" is a method of planning which cor-
relates educational specialists with specific position requirements
within the various sectors of the economy. There are four important
components to this method which require the collection of data on the
estimation of future needs.[1] Future requirements are estimated from
classified lists of posts to be filled by specialists; estimates of specialist
requirements and effects of the increase in positions; estimates of the
need for specialists to compensate for workers leaving their positions;
and estimates of the need for specialists to replace workers trained
other than through higher and specialized secondary educational institutions so as to upgrade these positions.

a. Classified Lists of Posts to Be Filled by Specialists

These lists consist of positions with their respective training requirements. The lists are compiled by all the economic units of the min-

---

[1]K. Nozhko, E. Monoszon, V. Zhamin, V. Severfsev, *Educational Planning in the U.S.S.R.*, UNESCO, 1968, pp. 139-145.

istries, from research units to producing enterprises. Each position identifies the degree of training required for individuals filling the various positions. The real question is whether the currently identified training levels will hold for the future. If not, then the projected type of requirements will be added to the list and the training necessary to supply such a person will be specified. Table IX-1 shows a general format for collecting and analyzing such data.

As a rule, these lists provide information on each type of post, as well as the quantity of engineers, technicians, etc., in various organizations and enterprises. With these data, ratios between specialists with higher education and those with specialized secondary education, as well as the total number of persons required, are calculated. From the lists and the ratios, future requirements for specialists are then estimated.

b. Estimates of the Requirements for Specialists

Making estimates of future requirements for specialists is complicated by the dynamics of the Soviet economy. Requirements in terms of both number and type of specialists are always changing. The major factors which are to be considered and tend to increase the requirement for specialists are (a) the establishment of new enterprises, institutes, organizations, design bureaus, etc.; (b) rising technological requirements in production which change the ratio of specialists to workers; and (c) reduction in the length of the working day. Those which tend to decrease the requirements are (a) improvements in the organization of the production unit; (b) mechanization; and (c) release of specialists from positions for which they are overtrained. From these data, the number of posts which need to be filled by specialists is calculated, including the number of teachers in the educational system.

c. Estimates of Needs for Specialists to Compensate for Workers Leaving Their Positions

These estimates include vacancies due to death, retirement, quitting for personal reasons, and positions currently held by persons not trained formally for those positions. This last category should decrease in importance as the number of persons with post-secondary training continues to expand. The number of persons in the various categories of "dropouts" differ from industry to industry and organization to organization. All these factors must be taken into consideration when

TABLE IX-1

LIST OF POSTS FOR SPECIALISTS WITH HIGHER OR SECONDARY SPECIALIZED EDUCATION

Structural subdivisions and posts	Total number of specialist posts	Of this number:			
		Posts to be filled by specialists with higher education		Posts to be filled by specialists with secondary education	
		Number	Listed specialization	Number	Listed specialization
Final result					
Ratio of specialists to total workers employed					

Source: K. Nozhko, E. Monoszon, V. Zhamin, V. Severfsev, Educational Planning in the U.S.S.R., UNESCO: International Institute for Educational Planning (1968).

determining the percentage to be utilized in long-term estimates for the loss of specialists with higher and specialized secondary education.

The effectiveness of the "nomenclature method" requirement for specialists is dependent upon integrating it into the five-year or annual plans. The method requires close correlation between five-year and annual plans and plans for development of higher and specialized education for scientists, engineers, and technicians. This method has the drawback that, with its requirement to be correlated with the five-year plan, it doesn't provide a sufficiently long time horizon to alter the flow of entrants into the various types and fields of education. This deficiency in the nomenclature method has resulted in its being utilized in conjunction with the saturation rate method.

2. Saturation Rate Method

Another method used for calculating the required number of specialists is the saturation rate method. The proper proportion of specialists to the total number of employed persons within any sector of the economy provides the data used to give the "saturation rate." This method requires the use of indices such as the planned volume of production and planned increases in labor productivity, presupposing knowledge of planned increases of employment in each sector. The coefficient of saturation (the ratio of specialists per 1,000 workers) is measured independently for each sector.

The saturation method is used for determining the manpower needs of all sectors of the economy. This covers not only the requirements of the upcoming planned period, but also the needs of subsequent periods, including projections for as long as fifteen years. Such long-term planning is necessary in order to determine the required levels of admissions into higher and specialized secondary educational institutions. The coefficient of saturation is applied to the calculated total number of workers to determine the required number of specialists. Each sector receives its own coefficient of saturation. It is assumed that the projected rate of increase of the total labor force will be representative of the increase in the specialized labor force; therefore, a calculation must be made for each sector of the economy. If only one coefficient of saturation were used for all sectors, it would artificially increase the requirements for some sectors and decrease them in others.

The main stages in applying this method are:

● Forecast of population;

● Forecast of the distribution of the work force among the main branches of the economy;

● Study of the number of specialists with secondary and higher education per one thousand workers in various sectors over time;

● Extrapolation by proportions, by branch, of staffing patterns to end of plan period;

● Calculations of the number of admissions to educational institutions to provide the stock of specialists necessary in the plan period;

● Calculations of the specific number by specialties of training plans, from the total estimates of the specialists needed.

The last step is particularly difficult because it requires that planners anticipate the necessary structural changes in the economy with their associated technological advances. These estimates must be more than simple extrapolations of historical trends. Soviet planners sometimes utilize staffing patterns of the more advanced or most rapidly advancing sectors to estimate future requirements for specialists in new areas for which major technological advances are foreseen.

The saturation rate method of making estimates of future needs for specialists and thus assisting in the planning of Soviet higher and specialized education must rely in part on the nomenclature method described above. The nomenclature method provides relatively accurate detailed data on the number of personnel in different specialities utilized in the Soviet economy. The two methods, when used simultaneously, can help verify the results of one another; with the help in recent years of the computer, they are generally utilized as complementary in deriving the final plans for scientists and engineers.

3. Extrapolation Methods

Extrapolation methods for the determination and forecasting of manpower needs are among the most popular and accessible in the USSR. A detailed discussion of this methodology is contained in "Methods for Forecasting the Number of Scientific Personnel" by F. V. Rossels.[2] The author states that the methodological basis of extrapolation methods is formed from two interconnected assumptions: the hypothesis that there are continuous smooth changes over the whole forecasting period and the assumption that some of the characteristics of the de-

[2]F. V. Rossels, "Methods for Forecasting the Numbers and Structure of Scientific Personnel," in D. M. Gvishiani et al., *The Scientific Intelligentsia in the U.S.S.R.*, Moscow, Progress Publishers, 1976, pp. 219-247.

velopment remain constant both in the base and the forecast period. Examples of the type of characteristics considered here would be the rate of growth of the number of scientific employees, and the acceleration of growth.

The question of errors in the forecast connected with a change in the trend observed in the base period is closely linked with determining the depth of the forecast, i.e., the length of the forecast period. The overwhelming majority of forecasters consider extrapolation to be suitable for quantitative calculations, for short- and sometimes medium-term forecasting, for up to approximately 5 to 15 years. Often, the depth of the forecast depends on the size of the base period. It is true that estimations on this vary noticeably from one forecast to another: some forecasters recommend taking a forecast period equal to one-third the length of the base period, others that they should be equal, and a third group that, in certain circumstances, the forecast period can exceed the length of the base period by several times. It will vary according to the kind of forecast being undertaken.

Rossels notes that forecasts obtained by mathematical extrapolation can produce good agreement between forecast and actual data, but the results cannot be directly used for planning and, in particular, for management. This is due to the fact that extrapolation is based on the statistical connections developed between parameters. Control of the way they function and develop requires a knowledge of the cause and effect relationships.

4. Complex Models of Forecasting

Soviet scientists have been developing more sophisticated models which attempt to overcome the limitations for planners of extrapolation models. Models constructed for forecasting the number of scientific personnel can be divided into three types: demand models, supply models, and balance models. These three types of models are also described in detail in the study by Rossels.

a. Models for Forecasting the Demand for Scientific Personnel

A priori, given values of aggregate parameters for the whole of the forecast period serve as the basis for almost all models of this type. The values of the aggregate parameters are generally set with the aid of extrapolation or of higher level forecasts, long-range plans, or central directives.

After selection of the aggregate parameters which serve as the basis of the forecast (e.g., the size of national income, the growth of labor productivity), the relationship between them and the number of science and engineering personnel is determined. Assumptions can then be made about the maintenance or changes in this relationship for that forecast period. These can be achieved through analysis, experimental estimations, analogies, or extrapolation. This process produces a system of equations which form a mathematical model. This model contains the known values of the aggregate parameters for specific moments of time, the link coefficients, and the unknown number of different scientists and engineers. The number of scientific personnel for a particular year can be forecast through the solution of the model.

b. Models of the Supply of Highly Qualified Specialists

Rossels notes that models have been developed in the USSR to estimate the future number of expected personnel in scientific fields in order to assist in the planning of the supply and training of such personnel. As the number potentially expected does not always equal the actual number obtained, some flexibility for retraining must be maintained.

"Input-output" models, which use mathematical formulations of the transfer process between professional levels, are considered effective forecasting tools. The system of scientific personnel is represented in the form of different blocks, or personnel qualification levels, and the connections between them. This determines the sequence of horizontal and vertical mobility of the personnel between these levels.

Figure IX-1 presents a block scheme of the amalgamated model developed by Rossels for the numbers of Soviet scientific workers (without distribution by speciality). Rossels made the following assumptions in the development of the scheme:

> 1. It was assumed impossible to make two or more transfers between levels within a single year (including jumping a level).
> 2. It was assumed that external postgraduates are considered by the Central Statistical Bureau of the USSR as scientific workers.
> 3. The possibility of mutual transfers between a group of scientific workers and persons not employed in the economy of the USSR (excluding students in higher education institutions) was not taken into account.[3]

This model also reflects that, according to current statistical reporting methods, persons with advanced degrees of Candidate or Doctor of

[3]Rossels, op. cit., p. 238.

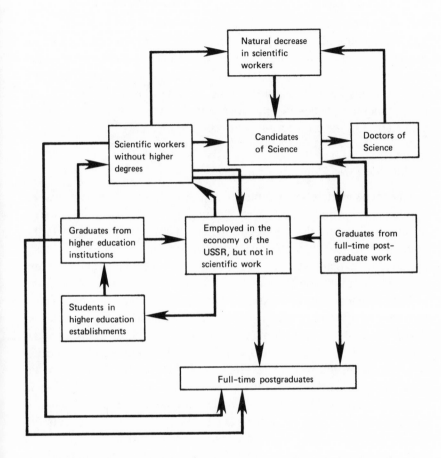

Source: D.M. Gvishiani, et. al., The Scientific Intelligentsia in the U.S.S.R., Moscow:
          Progress Publishers, p. 238 (1976).

Figure IX-1   AMALGAMATED MODEL FOR THE NUMBERS OF
                SOVIET SCIENTIFIC WORKERS

Science are considered as scientific workers regardless of the place and nature of their work.

The model follows the progression of the scientific worker from entrance (upon graduation from the higher educational institution, completion of postgraduate work, or through promotion of workers without degrees to Candidates and Doctors of Science) to exit from either retirement or death. Rossels notes that accurate quantitative calculations cannot yet be made on this type of model, as it utilizes only the most general parameters and does not yet have the necessary data characterizing the connections between the elements.

c. Balance Type Models

Balance type models are designed to achieve a specific balance between the demand for and supply of specialists. In these models, the exogenous variables are the initial parameters. These variables include the number of specialists who must be produced each year of the forecast period, and, dependent on these numbers, the acceptances and transfers required from one educational level to another.

Three separate models actually compose the balance model. These include (1) a demographic model of the movement of the population to determine the overall human resources potential (a supply model); (2) a demand model for specialists of different qualifications; and (3) a model of the movement of students and scientific workers in the educational system and in the fields of science. The third model is sometimes solved like a supply model and sometimes like a balance model. This is to ensure agreement between the data of models (1) and (2) by means of a series of consecutive iterations.

Rossels notes that the construction of a model of the demand for scientific personnel is extremely complex because no quantitative theory exists at the present time to link such parameters as gross national product, national income, and labor productivity with parameters characterizing the level of development of science (expenditures on scientific research, the numbers of scientific personnel, the number of scientific research institutes, inventions, publications, innovations, and so on).

Such a composite model for the calculation of the numbers and structure of scientific workers in the future has not yet been formulated because of the lack of the necessary statistical information and methodological difficulties connected with determining the demand for highly qualified specialists of different types. Rossels notes that the first obstacle is mainly organizational in nature and can probably be eliminated. However, the difficulties inherent in determining the demand in the economy for scientists are more general in nature, representing a question of the role of science in the contemporary economy and the efficiency of scientific workers.

5. Projections of the Supply of Scientists and Engineers in the USSR

The Soviet Union does not routinely publish projections of the supply of scientists and engineers. However, it is possible to make rough estimates of what the future supply might be.

Table IX-2 shows two estimates of the number of diplomas, by broad fields of study, granted to graduates of higher educational institutions in the Soviet Union. These represent high (Variant A) and low (Variant B) estimates for the 1980-1990 period. Variant A assumes that the number of graduates will grow at the rate experienced in the period 1975 to 1979.[4] These years were selected for the projections because the higher rates of increases in the sixties and early seventies were not sustained in the last half of the latter decade and the rate seemed to be stabilizing. Variant B uses the average percentage of the 23-year-old population graduated in each broad field in the 1975 to 1979 period, applied to population projections for 1980 to 1990.

Variant A, which shows a steady increase in the number of graduates, assumes that the Soviet leadership will increase the percentage of total graduations of the 23-year-old population from the average rate of 16.6 percent in the 1975 to 1979 period to 24.3 percent in 1990, with graduations in the physical and life sciences and mathematics rising from 1.0 percent to 1.6 percent and graduations in engineering rising from 6.4 to 10.0 percent of the 23-year-old population. The comparable rate for all fields combined experienced by the United States in the 1970s was between 24 and 25 percent of the 23-year-old population (See Chapter IV, Table IV-9). Variant B, which shows a decline in the number of graduations beginning about 1984, clearly reflects the prospects for the decline in this age group of the population in the Soviet Union in the 1980s.

The actual number of graduates is likely to be somewhere between the two estimates. Although the population changes will be a severe constraint on continued growth at the 1975-1979 rates, Soviet planners can increase the percentage of enrollment in the science and engineering fields in order partially to offset the demographic trends.

Projections of Candidate and Doctor degrees awarded are much more

[4]These projections were based on a least-squares linear regression, calculated for the 1975 to 1979 period.

TABLE IX-2

PROJECTIONS OF USSR GRADUATIONS FROM HIGHER EDUCATIONAL INSTITUTIONS: 1980 through 1990

(In Thousands)

Year	23-Year-Old Population	Total, All Fields		Graduations Total, Science & Engineering		Physical & Life Sciences & Mathematics		Engineering		Agriculture	
		Variant A	Variant B	Variant A	Variant B	Variant A	Variant B	Variant A	Variant B	Variant A	Variant B
1980	4871	809.3	807.6	430.3	424.8	51.3	50.7	316.9	310.8	62.1	61.4
1981	5000	828.3	829.0	441.9	436.0	52.6	52.0	325.8	319.0	63.5	63.0
1982	5090	847.3	843.9	453.5	443.8	53.8	52.9	334.7	324.7	65.0	64.1
1983	5169	866.3	857.0	465.1	450.7	55.1	53.8	343.6	329.8	66.5	65.1
1984	5151	885.4	854.0	476.8	449.2	56.3	53.6	352.5	328.6	67.9	64.9
1985	4964	904.4	823.0	488.4	432.9	57.6	51.6	361.4	316.7	69.4	62.5
1986	4752	923.4	787.9	500.0	414.4	58.8	49.4	370.4	303.2	70.9	59.9
1987	4512	942.4	748.1	511.7	393.4	60.1	46.9	379.3	287.9	72.3	56.0
1988	4249	961.4	704.5	523.3	370.5	61.3	44.2	388.2	271.1	73.8	53.5
1989	4123	980.4	683.6	534.9	359.5	62.6	42.9	397.1	263.0	75.3	51.9
1990	4041	999.4	670.0	546.6	352.4	63.8	42.0	406.0	257.8	76.8	50.9

Note: Variant A projections based on 1975 through 1979 data contained in Chapter IV, Table IV-9, using least-square linear regression; Variant B projections use the average percentage of the 23-year-old population in the respective fields for 1975 - 1979 as shown in Table IV-9.

Source: USSR population estimates for July 1 of each year, prepared in March 1977 by the Foreign Demographic Analysis Division, U.S. Bureau of the Census.

difficult to make because of the diversity of the means of pursuing the Candidate degree and the fact that the Doctor degree is really awarded for the research performance of the individual rather than for completing a prescribed course of study. Nevertheless, some order of magnitude projections can be presented to suggest possible trends in the number of persons receiving these degrees for benchmark years 1979, 1985, and 1990.

Projections of total Candidate and total Doctor degrees awarded through 1990 have been made by the Defense Intelligence Agency (DIA).[5] The base data used for the projections are the number of persons in formal aspirant training. As the number of aspirants remained relatively level in the 1972-1976 period, and there has been no evidence of Soviet planners' intent to change it, the average number of aspirants in the five-year period 1972-1976 was held constant throughout the projection period. The DIA estimates use this base number of aspirants, adjusted for time lags in getting the degree to make projections of total Candidate and total Doctor degrees awarded shown in Table IX-3. Rough estimates of breakdowns by broad field of science have been made by taking the percentage of graduate enrollment in these fields in 1974[6] and applying it to the total degrees awarded as projected by DIA.

As the methodology would lead one to expect, the number of Candidate degrees awarded remains relatively constant throughout the period, after dropping from the 1979 level. While the number of Doctor of Science degrees awarded rises throughout the projection period, given the approximate ten-year period of time the average Doctor of Science degree takes to be awarded, a decline could then be expected in later years. These data indicate, as does Variant B in Table IX-2, that demographic problems in the Soviet Union may force a reduction in the number of graduates at the higher education and Candidate levels in the 1980s and at the Doctorate level in the 1990s. The methodology does not, however, account for the possibility that Soviet planners may increase the level of enrollment in formal aspirant training in response to demographic constraints and as a result, produce a greater number of Candidates and Doctors of Science than the data would indicate. And, again, there is the possibility that Soviet planners may increase the percentage enrollment in the science and engineering fields in response to declining overall enrollment patterns.

[5]Jill E. Heuer, *Soviet Professional, Scientific, and Technical Manpower*, Defense Intelligence Agency, DST-1830 S-049-79 (May 1979), pp. 101-104.
[6]See Chapter V, Table V-1.

TABLE IX-3

PROJECTIONS OF USSR CANDIDATE OF SCIENCE AND DOCTOR OF
SCIENCE DEGREES AWARDED: 1979, 1985 and 1990
(In Thousands)

Candidate of Science Degrees	Total, All Fields	Total, Science & Engineering	Physical & Life Sciences & Mathematics	Engineering	Agriculture
1979	27.8	20.8	7.8	11.5	1.6
1985	26.3	19.7	7.4	10.8	1.5
1990	26.4	19.7	7.4	10.9	1.5
Doctor of Science Degrees					
1979	4.9	3.7	1.4	2.0	.3
1985	7.9	5.9	2.2	3.3	.4
1990	9.5	7.1	2.7	3.9	.5

Sources: Total Candidate and total Doctor degree projections from Soviet Professional, Scientific, and Technical Manpower, DIA, May 11, 1979, p. 99; breakdowns by field of science based on aspirant enrollment percentages for 1974, shown in Chapter V, Table V-1.

B. Projecting Future Supply and Demand for Scientists and Engineers in the United States

The supply and demand for scientists and engineers in the United States is forecast primarily to provide information for students, workers, the government, employers, and educational institutions participating in the market. Forecasts provide those entering a field of specialization an indication of their future chances of finding employment and their potential earnings. Both Federal and State governments have to determine levels of support for education, and need to know the implications of program changes on the availability of specialists. Employers require information to determine recruitment, training, utilization, salaries, and capital investment. Educational institutions provide both a supply and demand for trained workers and require information for long-range planning of construction and staffing and for instructional program offerings.

In the United States there are three basic methodologies for forecasting the market for scientists and engineers. Two are closely associated with government agencies—the Bureau of Labor Statistics (BLS) and the National Science Foundation (NSF)—and are best described as extrapolation and econometric models. The third model, pioneered by Richard B. Freeman, is a dynamic model.

1. Forecasting Scientific and Engineering Requirements and Supply (Bureau of Labor Statistics Projections)

The Bureau of Labor Statistics (BLS) utilizes two projections to analyze the differences between the supply of engineers and scientists and the requirements for this professional manpower. The supply component is drawn from data compiled by the National Center for Education Statistics (NCES) while BLS makes its own projections of requirements.

The National Center for Education Statistics generates the degree projections which the Bureau of Labor Statistics utilizes. NCES derives its estimates of future degrees by extrapolating past trends. The trends used in the projections are the ratios of total doctoral awards to the college-age population and the ratios of degrees in a field to total doctorates.[7]

The scientific and engineering requirement projections are somewhat more complex as they require several steps:

[7]National Science Board, *Science Indicators 1978*, p. 248 (1979).

a. Potential gross national product (GNP) is projected at close-to-capacity levels (assuming that 96 percent of the projected civilian labor force will be employed, that output per man-hour will grow at its long-term rate, and that average hours of work will decline slightly).

b. Projected GNP is distributed among personal consumption, investment, and government purchases of goods and services via a macroeconomic model in which the amount of each good and service finally consumed is estimated by patterns of consumer and government expenditure and of capital investment.

c. Production levels in each industry—including raw materials, transportation, components, and services—required to produce final products are estimated by input-output relationships among industries. Results of this method are checked against each industry's production estimates from a regression analysis using the industry's output and the major factors that have been found to affect it.

d. Projected industry production is converted into projected industry employment by projecting changes in productivity and in hours of work.

e. Projected requirements for workers in each occupation in the industry are estimated from the occupational composition adjusted to reflect expected changes in technology and patterns of work organization in the industry. The numbers of scientists and engineers engaged in R&D in each industry are checked against projections of R&D expenditures. The projections for the major industries are discussed with management in order to attain information on market, technological, or other factors that have impinged upon past output and employment trends and that may affect future output and employment.

f. Projection methodology is different for higher education and government employing scientists and engineers. Programs composing Federal, State, and local government employments—schools, health, highways, defense, police—are projected separately because each makes different use of different occupations. For government employment, projections are essentially guesses about legislative appropriations and are not based on market behavior.

g. The total requirements for each occupation are then summarized. To the net growth in requirements in each occupation is added an estimate of the replacement needs resulting from deaths and retirements based on the age composition of the members of the occupation.[8]

These projections assume a continuation of present economic patterns, of the growth rates of different sectors, and of utilization rates for different occupations independent of the supply. Projections of the manpower supply are also based on the continuance of present patterns of occupational choice. As this prevents accurate forecasting of demand and supply balancing by the wage mechanisms, two projections are made to illustrate divergences between anticipated supply and demand for manpower. These give an indication of the direction and magnitude of potential imbalances.

2. Method of Forecasting Science and Engineering Ph.D. Requirements and Supply (National Science Foundation Projections)

The employment outlook for science and engineering Ph.D. holders has been an area of particular interest. The National Science Foundation has collected basic manpower data on a regular time series and provides projections of scientist and engineer supplies and utilization. These data include Ph.D. holders in engineering, mathematics, natural sciences, and social sciences.

Differences in the methodologies utilized by the Bureau of Labor Statistics and the National Science Foundation have been described as follows:

In contrast [to BLS projections], NSF uses econometric modeling and trend extrapolation to estimate the number of science-and-engineering-related positions which may be available by field for doctorates in selected years. This concept of utilization is based on the type of activities in which doctorates are engaged. NSF projects the two largest categories of science and engineering employment, academic and industrial R&D, through the use of demand equations that are derived from regression analysis. Demand variables include R&D spending and the number of science and engineering baccalaureates awarded (an index of teaching loads) in a year. Other categories of science and engineering employment are projected through extrapolation.[9]

Projections of scientific and engineering manpower assume a set of

[8]Joel L. Barries, "Projecting Future Supply and Demand for Scientists, Engineers and Technicians," in Training and Utilization of Scientific and Engineering-Technical Personnel, SRI International, pp. 97-99 (November 1979).
[9]Science Indicators 1978, p. 248.

national environments for the period of projection. The environment for the supply and utilization of science and engineering doctorates is determined by several general factors, including the economic climate of the country, the nature of the higher educational system, the working life patterns of the labor force, and the relative international economic and technological position of the United States.[10] Several key economic indicators provide levels and rates of growth of an economy, and serve as the direct basis for the projections of utilization and the indirect basis for projections of supply.

Other aspects of the environment implicitly assumed by NSF in projections of manpower supply and utilization are:

- The institutional framework of the economy will not change significantly within the projected period, and the role of the labor force will follow past trends.
- There will not be significant reductions in defense expenditures.
- The role of science and technology is expected to become more important to the operation of programs dealing with national, regional and local problems.[11]

Inherent in the NSF projections are some basic premises that tend either to encourage or discourage the production of and expanded demand for doctorates. Expansionary assumptions include:

- An "oversupply" of doctorates adjusts doctorate salaries downward relative to nondoctorate salaries.
- The doctorate degree constitutes a "ticket" to a frequently preferred professional or academic life and work style (regardless of economic considerations). This phenomenon is likely to continue.
- The doctorate degree may become a prerequisite for positions currently being filled by nondoctorates, in part because of the availability of doctorates and in part because of the positions' increasing technical content.[12]

Contractionary assumptions affecting the production of and expanded demand for doctorates are:

- In the early seventies, proportionately fewer college-age persons

---

[10]Barries, *op. cit.,* p. 102.
[11]Barries, *op. cit.,* pp. 102-104.
[12]Ibid., p. 104.

have entered college, possibly because of the reduction of job opportunities and the fall in relative salaries of college graduates.

• College students will be discouraged from continuing their education to the doctorate level if: (1) the expected reduced growth (in comparison to the sixties) in the demand for college faculty and researchers continues; and (2) the relative earnings of doctorate degree holders move toward that of master's and bachelor's degree holders.

• Increases in tuition and reduction in graduate student support will reduce the number of students pursuing a doctorate degree.[13]

Each set of the above-mentioned factors is expected to moderate the effects of the other; no particular set of factors is expected to dominate.

Two sets of models have been developed by NSF for both supply and utilization, and are labeled "Probable" and "Static." Probable models project what is thought to be the more likely course for future events while Static models provide benchmarks for comparison and illustrate the effects of the continuation of past and current trends. Trends for the two models are consistent, despite different numerical results.

These models (BLS and NSF) assume continuation of present economic patterns. The models can be redesigned and adjusted to make projections for sectors or industries in which major changes in technologies or national priorities have occurred (for instance, the impact of computers or the attempt to analyze the variety of scenarios concerning energy policies and their impact on scientific and engineering manpower).

3. Dynamic Models

The preceding sections on U.S. forecasting have covered the basic methodologies of science and engineering manpower and utilization projections made by U.S. agencies. The NSF models have focused on Ph.D.s, while the BLS methodology has been employed as much or more at the bachelor's level. These techniques generally extrapolate past trends without taking into account socioeconomic factors which may have an impact on a career decision. Future levels of supply and demand are separately projected: comparison of the two gives estimates of expected future manpower imbalances. Future levels of supply are generally projected assuming a constant or predictable rate of school

---

[13]Barries, *op. cit.*, p. 104.

attendance, while demand is projected for different levels of anticipated GNP assuming a constant relative output mix, utilizing input/output coefficients to determine science and engineering manpower needs. These input/output coefficients can be adjusted to account for technological changes that may occur.

Extrapolation or fixed coefficient projection models assume that either labor markets are relatively consistent or that economic conditions are irrelevant in career decisions. This approach has been criticized as it does not take into account the adjustment mechanisms of the labor and product markets. These models are also criticized as they do not include relative wage and price levels, do not account for supply behavior and the interactions and feedback among economic variables, and are generally disassociated from policy.

A market projection model has been developed as an alternative, which includes the effects of labor market mechanisms. The first is the human capital theoretical approach to career decisions, outlined in Chapter I. Career decisions are made to maximize expected net gain. This involves explicit or implicit projections of future expected total compensation, both monetary and nonmonetary, of all possible career choices. This is compared to the required costs of attainment. The career choice is made, subject to ability and imperfect capital market constraints, on the basis of the highest expected discounted net gain. The decision is also subject to a risk premium.

Application of this theoretical structure as a testable model capable of prediction presents difficulties. The most successful construction and application of models has been performed by Richard Freeman.[14] Current market salary is used in the specific field while an average current market salary is constructed as the alternative for all other options. The current and future supply of new entrants into the market are predetermined four years in advance of entry, the amount of time required to finish the educational process, and are thus linked to past supply and demand levels. The demand for scientific and engineering personnel and the number supplied by graduations determine current salary levels. Demand for scientific and engineering personnel is determined by levels of R&D and durable goods expenditures. As supply is predetermined by four years and as the salary levels are based on

[14]Richard B. Freeman, *The Market for College-Trained Manpower: A Study in the Economics of Career Choice*, Harvard University Press (1971); "Supply and Salary Adjustments to the Changing Science Manpower Market: Physics, 1948-1973," *American Economic Review*, Vol. 65, No. 1, pp. 27-39 (March 1975); "A Cobweb Model of the Supply and Starting Salary for New Engineers," *Industrial and Labor Relations Review*, Vol. 29, No. 2, pp. 236-248 (January 1976); *The Over-Educated American*, New York: Academic Press (1976).

current levels of supply and demand, a "cobweb" model is formed, whose recursive structure produces "endogenous cyclical fluctuations," which are independent of economy-wide exogenous fluctuations.

The model designed for estimating the supply and starting salary of new engineers consists of the following four equations:[15]

(1) Supply of New Entrants
     $ENT = a_1 \ SAL^* \ (0) - a_2 \ ASAL^* \ (0) + \mu_1$

(2) Supply of Graduates
     $GRAD = b_1 \ ENT \ (-4) + b_2 \ [SAL(-3) + SAL(-2)]$
     $- b_3 \ [ASAL(-3) + ASAL(-2)] + \mu_2$

(3) Salary Determination
     $SAL = C_1 \ RD + C_2 \ DUR - C_3 \ GRAD + \mu_3$

(4) Salary Expectations
     (a) $SAL^* + SAL; \quad ASAL^* = ASAL$
     (b) $SAL^* = \lambda \ SAL + [(1-\lambda) \ SAL^*(-1)];$
          $ASAL^* = \lambda \ ASAL + [(1-\lambda) \ ASAL^*(-1)]$

Where

ENT	=	first-year enrollments in engineering
SAL^*	=	expected engineering salaries in the market four or more years in the future
$ASAL^*$	=	expected alternative salaries in the market four or more years in the future
SAL	=	actual engineering salaries
ASAL	=	actual alternative salaries
GRAD	=	number of engineering graduates
RD	=	research and development spending
DUR	=	durable goods output

Equation (1) relates the enrollment decision to expected engineering and alternative salaries. Equation (2) makes the number of graduates four years later a function of first-year enrollments and additional salary information received by students during the first two years of study, on the assumption that changes in majors occur prior to the junior year. The demand side of the market is represented by equation (3), which links salaries to the number of graduates and two exogenous demand-shift variables: R&D spending and durable goods production. Finally, equation (4) presents two alternative hypotheses concerning the relationship between salary expectations and measurable variables: (a) expected salaries equal current salaries and (b) expected salaries are formed by adaptive expectations and equal some combination of current salaries and past expectations. The latter hypothesis yields a cobweb theory.

[15]Freeman, "A Cobweb Model," *op. cit.*, pp. 241-242.

Freeman's model is used primarily for determination of the numbers and salary level of entering science and engineering manpower. The model can be applied to an entire field, although some conceptual and estimation difficulties remain. Experienced and inexperienced scientists and engineers are not perfect substitutes even for a narrow field: experienced workers have often attained a high level of on-the-job training and are often employed in different occupations within a field.[16]

This model has been relatively successful in explaining variations in starting salaries and numbers of new entrants in some fields of engineering. However, Freeman's approach did not outperform the fixed coefficient model in projecting changes in employment among industries. Several additional criticisms have been made, however, about its effectiveness in predicting future market behavior:

• The very broad definition of scientists and engineers does not allow forecasts of crucial subfields in which market imbalances can occur;
• The total number of students entering college should be an endogenous variable instead of an exogenous variable;
• The use of current salaries as proxies for expected lifetime discounted total compensation will create inconsistent projections unless the ratio of salaries to lifetime discounted total compensation are constant among all fields and in all time periods;
• The simultaneous estimation of individual labor market models when placed in a system creates problems not encountered in single-labor market estimations;
• The models do not allow for entry to and exit from a field, and consequently do not project mobility across fields that may occur for economic or technical reasons.

The broadness of the definitions of fields lessens this last problem to some extent, as is seen in Chapter VIII, as most mobility is within technologically related fields.

It is generally felt that models for forecasting the supply of and demand for scientists and engineers require much more development before they can be realistically used for projections. However, they are the beginning of a potentially more useful projection methodology and will be used more in the future as they are further developed.

---

[16]Richard B. Freeman ''An Empirical Analysis of the Fixed Coefficient 'Manpower Requirements' Model, 1960-1970,'' *Journal of Human Resources*, Vol. 15, No. 2, Spring 1980, p. 176-199.

4. Projections of the Supply of Scientists and Engineers in the United States

The basic data for both the NSF and BLS projections are those compiled by the National Center for Education Statistics. The National Center for Education Statistics makes its own projections of enrollment and earned degrees at the bachelor's, master's, and doctoral levels.[17] These projections utilize a basic methodology, which calculates the rates for recent years as a percentage of a "base" variable, for example, enrollment for a given age group as a percentage of the population of that age group for each of the last 10 years. The rate is then projected into the future and applied to projections of the "base" variables. The following projections in Tables IX-4, IX-5, and IX-6 are dependent upon the enrollment projections.[18]

Table IX-4 shows projected earned bachelor's degrees by field from 1977-78 through 1988-89. The number of engineering degrees projected to be granted rises into 1981-82 and then subsides. Degrees awarded in the physical and life sciences follow a similar pattern of rising in the early 1980s and then falling.

The National Center for Education Statistics has also projected earned master's degrees to 1988-89, as shown in Table IX-5. Although the projected numbers of degrees awarded in all fields decreases from 1983 throughout the remainder of the projection period, the number of degrees projected in the science and engineering fields remains relatively stable throughout the period with the exception of a general downward trend in the 1980s in the engineering fields. Data for earned doctoral degrees by field, shown in Table IX-6, show a gradual but steady decline in projected degrees throughout the 1980s, with a greater percentage decrease in the science and engineering fields than in all fields combined. Projected doctoral degrees awarded in the physical and life sciences and mathematics decline by almost 19 percent.

The Bureau of Labor Statistics (BLS) has made projections of new graduates at the bachelor's, master's, and doctoral levels, by specific occupation.[19] These BLS supply projections rely upon the National Center for Education Statistics estimates. BLS has also made estimates of occupational requirements. The training data are independent estimates but, when compared with projected occupational employment needs, reveal the possible imbalances between supply and demand.

---

[17]*Projections of Education Statistics to 1988-89*, National Center for Education Statistics, 1980.
[18]For a complete methodology, see ibid., pp. 139-151.
[19]*Occupational Projections and Training Data*, 1980 Edition, U.S. Department of Labor, Bureau of Labor Statistics, Bulletin 2052 (September 1980).

TABLE IX-4

PROJECTED U.S. EARNED BACHELOR'S DEGREES, BY MAJOR
FIELD OF STUDY: 1978-79 TO 1988-89
(IN THOUSANDS)

Year	Total, All Fields [1]	Total, Science & Engineering	Physical & Life Sciences and Mathematics	Engineering	Agriculture
1978-79	1,001.2	184.9	94.7	67.0	23.2
1979-80	1,017.3	194.5	96.1	73.8	24.6
1980-81	1,021.5	201.4	97.4	78.3	25.7
1981-82	1,035.8	213.0	100.0	86.6	26.4
1982-83	1,027.9	214.7	101.6	86.0	27.1
1983-84	1,014.3	213.2	101.0	84.6	27.6
1984-85	994.2	210.1	99.5	82.8	27.8
1985-86	991.9	207.8	99.1	80.8	27.9
1986-87	970.7	204.1	97.5	78.6	28.0
1987-88	967.0	202.8	97.5	77.2	28.1
1988-89	966.4	203.0	98.1	76.3	28.6

[1] Includes first professional degrees.

Note: Data are for the 50 States and the District of Columbia.

Source: National Center for Education Statistics, Projections of Education Statistics to 1988-89, April 1980, pp. 62, 66, 67 and 94

TABLE IX-5

PROJECTED U.S. EARNED MASTER'S DEGREES BY MAJOR
FIELD OF STUDY: 1978-79 TO 1988-89
(IN THOUSANDS)

Year	Total, All Fields	Total, Science & Engineering	Physical & Life Sciences and Mathematics	Engineering	Agriculture
1978-79	314.3	39.5	18.9	16.5	4.1
1979-80	315.1	39.1	18.3	16.6	4.2
1980-81	315.9	39.5	18.7	16.4	4.4
1981-82	315.9	40.1	19.2	16.3	4.6
1982-83	316.3	40.8	19.6	16.4	4.8
1983-84	313.2	41.0	19.8	16.3	4.9
1984-85	313.2	41.4	20.1	16.3	5.0
1985-86	311.9	41.2	19.9	16.2	5.1
1986-87	310.5	41.1	19.8	16.2	5.1
1987-88	301.0	40.5	19.3	16.1	5.1
1988-89	295.4	40.1	19.0	16.0	5.1

Source: National Center for Education Statistics, Projections of Education Statistics to 1988-89, April 1980, pp. 62, 72 and 73.

TABLE IX-6

PROJECTED U.S. EARNED DOCTORAL DEGREES BY MAJOR
FIELD OF STUDY: 1978-79 TO 1988-89
(IN THOUSANDS)

Year	Total, All Fields	Total, Science & Engineering	Physical & Life Sciences and Mathematics	Engineering	Agriculture
1978-79	32.0	10.6	7.3	2.3	1.0
1979-80	32.8	10.9	7.5	2.3	1.1
1980-81	33.0	10.7	7.3	2.3	1.1
1981-82	32.7	10.6	7.2	2.3	1.1
1982-83	31.7	10.3	6.9	2.3	1.1
1983-84	31.0	10.1	6.8	2.2	1.1
1984-85	30.4	9.9	6.6	2.2	1.1
1985-86	29.8	9.6	6.4	2.1	1.1
1986-87	29.2	9.5	6.3	2.1	1.1
1987-88	28.6	9.3	6.1	2.1	1.1
1988-89	28.0	9.0	5.9	2.0	1.1

Source: National Center for Education Statistics, Projection of Education Statistics to
1988-89, April 1980, pp. 62, 78 and 79.

However, the estimates do not consider any potential responses that could occur due to changing wage rates which might help equate supply and demand. The BLS projections contribute to the analysis of future occupational openings, but these estimates are sensitive to the assumptions of the BLS model and do not include occupational transfers. Only the BLS supply projections will be presented here.

Table IX-7 shows the BLS projections for the annual average number of degrees awarded for 1978 through 1990 for selected fields. It is important to note that significant numbers of persons earning degrees in a field may not enter that field—for instance only about 80 to 84 percent of the graduates with a bachelor's degree in engineering take engineering positions. Some go on for advanced engineering degrees, while others take positions in related fields.

As noted above, the National Science Foundation has focused its projections on doctoral scientists and engineers. Table IX-8 shows the full-time doctoral science and engineering labor force by field for 1979 and projected for 1990.[20] The data show the total science and engineering doctoral labor force, those employed in science and engineering and those with science and engineering degrees but employed in non-science/engineering fields. It is important to note that the percentage of doctoral scientists and engineers not employed in occupations for which they were trained is projected to rise from 9 percent to 18 percent between 1979 and 1990.

Although the NSF projections generally indicate that the number of Ph.D.s awarded in science and engineering will be sufficient to fill the demand for them, there is now evidence which seems to suggest that the number of Ph.D. engineers will fall short of demand, because many students are terminating their education at the bachelor's level to take advantage of strong demand (manifested in high salaries) for bachelor's degree holders in engineering. A recent study which cites the NSF estimates suggests that because the NSF projections did not account for the continued strong demand for new B.S. engineers to enter the labor force, future graduate engineering enrollments and, thus, future doctoral supply, may be overestimated.

It should also be noted that while the number of Ph.D.s in engineering granted by U.S. universities between 1974 and 1980 dropped by almost 20 percent, an increasing number of these degrees are being

[20]These data prepared by NSF were published in *Science and Engineering Education for the 1980's and Beyond*, National Science Foundation and Department of Education, (October 1980). The projections represent an update of the Foundation's "Projections of Science and Engineering Doctorate Supply and Utilization: 1982 and 1987," NSF 79-303.

TABLE IX-7

PROJECTED U.S. ANNUAL AVERAGE NUMBER OF NEW GRADUATES
AT THE BACHELOR'S, MASTER'S AND DOCTORAL LEVELS,
FOR SELECTED FIELDS: 1978 TO 1990

	Actual 1977-78	Projected Annual Average 1978-90
Engineers (Including Engineering Technology)		
Bachelor's Degrees	56,009	81,441
Master's Degrees	16,409	16,722
Doctor's Degrees	2,440	3,158
Life Science Occupations		
Biochemists		
Bachelor's Degrees	1,752	2,275
Master's Degrees	319	374
Doctor's Degrees	429	478
Life Scientists		
Bachelor's Degrees	74,937	88,268
Master's Degrees	10,887	13,972
Doctor's Degrees	4,284	4,096
Mathematics Occupations		
Mathematicians		
Bachelor's Degrees	11,886	11,616
Master's Degrees	2,640	2,667
Doctor's Degrees	592	508
Statisticians		
Bachelor's Degrees	273	300
Master's Degrees	507	510
Doctor's Degrees	153	131
Physical Scientists		
Chemists		
Bachelor's Degrees	11,474	12,523
Master's Degrees	1,892	1,483
Doctor's Degrees	1,525	1,199
Physicists		
Bachelor's Degrees	3,259	3,290
Master's Degrees	1,270	1,204
Doctor's Degrees	841	763

Source: Occupational Projections and Training Data, 1980 Edition. U.S.
Department of Labor, Bureau of Labor Statistics, Bulletin 2052,
September 1980; pp. 55-60.

TABLE IX-8

FULL-TIME U.S. DOCTORAL SCIENCE AND ENGINEERING LABOR FORCE, BY FIELD:
1979 ACTUAL AND 1990 PROJECTED

	Physical Sciences	Engineering	Mathematical Sciences	Life Sciences	Social Sciences	Total
Labor Force (In Thousands)						
1979	73	49	21	79	83	306
1990	103	80	30	113	125	450
Science/Engineering Utilization (In Thousands)						
1979	67	47	20	74	71	278
1990	93	63	23	93	99	370
Non-Science/Engineering Utilization (In Thousands)						
1979	7	3	1	5	12	28
1990	10	17	7	20	26	80
Non-Science/Engineering Utilization (Percent of Labor Force)						
1979	9%	6	6	6	14	9
1990	10%	21	23	18	21	18

Source: Science and Engineering Education for the 1980's and Beyond, prepared by the National Science Foundation and the Department of Education, October 1980, p. 31.

awarded to foreign nationals. (See Figure IX-2.)[21] In 1974, about 39 percent of Ph.D.s in engineering were granted to foreign students, while in 1980, foreign students accounted for about 46 percent of these degrees. At the same time, many of these foreign nationals appear to be obtaining permanent employment in the U.S. labor market. The U.S. Department of Labor reports that the number of permanent labor certificates for engineering occupations issued to foreigners more than doubled from 1976 to 1980. Thus, it appears that foreigners entering or remaining permanently in the U.S. labor market may be partially offsetting a shortfall in Ph.D.s in engineering.

C. Summary

Soviet and U.S. forecasters use many of the same statistical tools in their forecasts of future numbers of scientists and engineers. However, their objectives and their models are quite different. Neither country has developed refined forecasts capable of producing accurate predictions of the full range of scientists and engineers. The consequence of this, however, seems to weigh heavier on the Soviet Union, where there is a command economy and where effective planning based on accurate information and reliable forecasts is essential.

In the U.S. system, imbalance in the supply and demand for scientists and engineers will be corrected by the market process, although the corrective period may be longer than many desire. The corrective process is initiated by the condition of the imbalance itself and although governmental policy may accelerate the correction, it is usually initiated by the market.

The Soviet Union faces a far more complex problem due to the nature of the planned system. Imbalances which occur in the system are generally the result of ineffective planning or ineffectual implementation of the plan. Thus, corrections require direct action by the planning apparatus. Current forecasting methods in the Soviet Union seems inadequate to meet the needs of the planners. The "nomenclature" and the "saturation rate" methods are, at best, information generating methodologies which fail to effectively link the supply and demand functions in the USSR. These two methods have been severely criticized by Soviet economists, who frequently report evidence of imbalances and misdirected human resources. In some cases there is

---

[21]"Foreigners Snap Up the High-Tech Jobs," *New York Times*, Sunday, July 5, 1981, p. F13, citing data from the National Science Foundation and the U.S. Department of Labor.

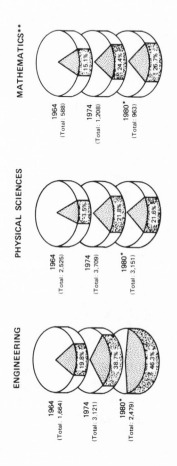

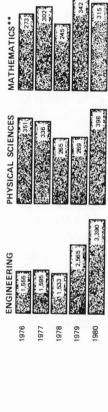

NUMBER OF PERMANENT LABOR CERTIFICATIONS ISSUES TO FOREIGNERS BY THE LABOR DEPARTMENT

*Preliminary **Including computer science

Sources: National Science Foundation and United States Department of Labor, as cited in "Foreigners Snap Up the High–Tech Jobs". New York Times July 5, 1981, p. F-13.

Figure IX-2   TOTAL NUMBER OF Ph.D.'s GRANTED BY UNITED STATES UNIVERSITIES AND THE PERCENT GOING TO FOREIGN STUDENTS

evidence of overshooting the planned ratio.[22] In other writings, even the objectives of the planners are called into question. It is asked, for instance, why the Soviet economy requires 2.7 times as many engineers as the United States, and why there are not more people in vocational and technical training.

The more sophisticated Soviet methodologies discussed above are far from being perfected and even further from full adoption and utilization in the planning process. Perhaps the most noteworthy aspect of the Soviet system is that it functions as effectively as it does in the *absence* of appropriate forecasting models.

The projections of U.S. and Soviet scientists and engineers in this chapter are based on assumptions that the forces (market and plan) operative during the 1970s will spill over into the 1980s. There are several observations that can be drawn from the U.S. and USSR historic and projected data of graduates. First, the Soviet Union is about to enter a period in which demographic factors will impose a constraint on its ability to continue to increase the stream of new science and engineering entrants into the labor force. This fact may help explain why the Soviet Union seems to have been over-producing engineers in the 1960s and 1970s relative to the ability of the system to absorb them efficiently: they may have been building human resource inventories on which they could draw as their demographic constraints tightened. The "over-supply hypothesis" could also be simply a reflection of the nature of the system to stockpile in advance of requirements to ensure the availability of supplies when they are needed.

In the next decade, the United States does not face as dramatic a demographic constraint as the Soviet Union. There is a significant reservoir of young people who, with appropriate incentives, could be redirected into the scientific and engineering fields. A major concern raised by the data is the decline in the engineering fields of Ph.D.s, who accept positions at colleges and universities as professors. However, the most evident constraint seems to be the failure of the leadership in the United States to grasp the necessity for a stated national policy for development of scientific and engineering manpower and the support of that policy with a commitment of resources. In the absence of any dramatic market shifts in the United States or strong commitment by the Federal government to provide incentives for better-trained and more scientists and engineers, the United States is not likely significantly to alter its comparative position with the USSR in these critical manpower areas.

---

[22]V. Gurin, *Sotsialistichesky trud*, pp. 152-155, May 1973.

CHAPTER X

Summary and Conclusions

This study has attempted to describe some of the general characteristics of the system of training and utilization of scientists and engineers in the USSR in contrast to that in the United States. Quantitative comparisons of the number of students enrolled in science and engineering programs at various levels of the educational process and of scientists and engineers employed in the economy as a whole and engaged in R&D have also been presented. This concluding chapter will summarize some of the main trends in U.S. and Soviet education and employment of scientists and engineers that have emerged in the preceding chapters by placing these findings within the context of the broad objectives of the Soviet Union for scientific and technical manpower.

A. Soviet Scientific and Technical Manpower Objectives

The Soviet leadership has long regarded the achievement of scientific and technological superiority not only as a major goal in itself but also as a prerequisite for the attainment of other strengths and capabilities to increase the chance of Soviet success in its competition with the West. Science and technology are viewed as important instruments in the struggle between the socialist and the capitalist systems.[1]

During the years following the Second World War, the Soviet Union experienced an impressive rate of economic growth as well as scientific and technological advance. Exhilarated by its space exploits and the high rate of technological achievement and economic growth experienced during the 1950s, the Soviet leadership was confident of its

[1]Soviet science and technology objectives are discussed in Francis W. Rushing and Catherine P. Ailes, "An Assessment of the USSR-US Scientific and Technical Exchange Program," *Soviet Economy in a Time of Change*, a Compendium of Papers submitted to the Joint Economic Committee, Congress of the United States (October 10, 1979), and Catherine P. Ailes, James E. Cole and Charles H. Movit, "Soviet Economic Problems and Technological Opportunities," *Comparative Strategy*, Vol. 1. No. 4 (1979).

ability to broaden its efforts to achieve world preeminence in science and technology and thereby enhance its power and influence abroad. By the mid-to-late-sixties, however, the leadership was increasingly aware that the Soviet Union had not been able to sustain its rapid rate of technological achievement and economic growth and was, contrary to earlier optimistic expectations, falling further behind the United States in the general development and utilization of science and technology in the economy as a whole, rather than narrowing the gap.

Economic growth in the fifties was principally a matter of quantitative growth in the inputs into production of primary factors— labor and capital, as well as cheap reserves of energy and raw materials. In a sense, even the very real achievements in a number of high-visibility scientific and technological areas were made possible by the Soviet ability to direct large resources to those areas. By the mid-to-late-sixties, it had become evident that the sources that the Soviet Union had previously relied upon as inputs for "extensive" economic growth were diminishing and would not provide the needed basis for maintaining growth of productive output at a rate comparable to that experienced in previous years.

As early as 1969, the slowdown in the growth of labor supply and other economic inputs had persuaded the Soviet leadership of the necessity of turning to more "intensive" forms of economic development than had been the case in the past. Brezhnev, at the December 1969 plenum of the Central Committee, stated that:

> Henceforth we must rely above all on *qualitative* factors of economic growth, on the heightening of the effectiveness [and] intensiveness of the national economy....This is becoming not only the main but also the only possible means of developing our economy and solving such fundamental socioeconomic tasks as the construction of the material, technical base of communism, the improvement of the welfare of the workers, and victory in the economic competition of the two world systems....*We have no other way*. (Emphasis added.)[2]

In formulating the Ninth Five-Year Plan (1971-1975), Soviet leaders opted for a major program of importing advanced technology and equipment from the West as a source of gain in productivity and as a solution to the economic deficiencies and management problems evident from the past record. However, apart from placing increased emphasis on the acquisition of Western scientific and technological

[2]L. I. Brezhnev, *On the Basic Problems of the Economic Policy of the CPSU at the Contemporary Stage* (Moscow: Politizdat, 1975), Vol. 1, pp. 418–419. (In Russian)

advances, Soviet doctrine and policy statements began to emphasize increased allocation to domestic research and development and to raising the qualifications of the labor force. In short, if the state imports technology, it must train people to absorb and exploit it.

In the development of the Tenth Five-Year Plan (1976-1980), intensification of economic production was the dominant theme. The plan, which was referred to as the plan of "effectiveness and quality," was based primarily on "the acceleration of scientific-technical progress, the rise in labor productivity, and the improvement in the quality of work by all possible means in all sectors of the national economy." The emphasis on intensification of production is evident throughout the plan, especially in statements of objectives for the five-year period, including that of raising the qualifications and professional skills of workers. The theme of intensification of the economy remains prominent in the draft of the Eleventh Five-Year Plan (1981-1985).

B. U.S.-Soviet Quantitative Comparisons

As indicated above, the clearly enunciated approach of the Soviet leadership for meeting the long-term future is firmly based on the application of the achievements of science and technology to all areas of the development of the bases of state power. An integral part of this doctrine is the restructuring of the skill levels of the labor force— a qualitative approach which is intended to provide the most effective avenue to quantitative improvements in the economic sphere. The Soviet Union has made remarkable strides in its goal of raising the quality of its work force to bring about the scientific and technological transformation of the economy. The Soviet commitment to the restructuring of the skill levels of the labor force emerges clearly from U.S.-USSR quantitative comparisons of the number of students enrolled in and completing science and engineering programs at various levels of the educational process as well as trends in the number of scientists and engineers employed in the economy as a whole and engaged in R&D in the United States and the USSR. Although caution should be exercised in drawing too many inferences from such quantitative comparisons, as differences in definitions and concepts used in the two reporting systems are quite marked, the following general trends can be observed:

● The level of educational attainment of the Soviet population as measured by number of years of training completed has risen from 5.8 years in 1960 to 8.7 years in 1977, and is projected to rise to

9.9 years by 1985. When these data are compared to data for the United States, there is still a significant disparity, especially in terms of percentage of population who have completed higher education (15.4 percent of total population in the United States with four or more years of college as opposed to 6.7 percent in the Soviet Union having completed approximately 5 years of higher education). The distribution of the Soviet population at the higher levels of education, however, is far more heavily skewed toward the scientific and technical fields than is that of the United States, and even at the lower levels of education there is a much greater emphasis on the science and mathematics content of the curriculum than is the case in the United States.

• U.S. elementary school students (grades K-6) receive slightly more hours per week in science training than do Soviet elementary school students, but the total hours devoted to mathematics at these grades is considerably below the average for the Soviet Union.

• The Soviet general secondary school curriculum is quite accelerated in both science and mathematics as compared with most curricula in U.S. high schools. The entire school population is exposed to the mathematics-science oriented curriculum in Soviet secondary schools rather than only selected students as is the case in the United States. Thus, in general, the Soviet secondary school graduate has a better training in mathematics and science than does his or her U.S. counterpart.

• The Soviet Union has developed a type of education—the specialized secondary school—that provides applied technical training at the secondary education level and is the principal source of the technical cadre that works under the direction of graduates of higher education, particularly in the engineering fields. Over one million students are graduated from specialized secondary schools annually. There is no broad-scale equivalent type of training in the United States.

• While the United States has a greater proportion of secondary school graduates entering higher educational institutions than does the Soviet Union, attrition rates are much higher. Over 80 percent of those admitted to higher educational establishments in the Soviet Union complete their undergraduate education and receive a diploma, whereas only about 55 percent of those students who enroll in U.S. colleges go on to receive their bachelor's degrees.

• As a percentage of college age population in 1978, the United States had nearly three times the number of students enrolled at higher educational institutions as the Soviet Union, and almost one-

and-one-half times as many total graduations of the 22/23-year-old population. However, in the engineering fields, the Soviet Union graduated almost six times the number of specialists at the under-graduate level as did the United States, a difference which is sub-stantial, even allowing for the probably inferior instruction of ap-proximately one-third of the Soviet engineering graduates who were enrolled in part-time programs.

• About 70 percent of Soviet graduate student (aspirant) enrollment is in the science and engineering fields. The percentage of total U.S. master's and doctoral enrollment in these fields has been steadily declining since 1960, and was about 20 percent in 1976 compared to 30 percent in 1960.

• At the end of 1974 (the latest year for which detailed breakdowns by field are available for the USSR), the Soviet Union and the United States had approximately the same number of specialists with ad-vanced degrees (Candidate of Science or Doctor of Science in the USSR and doctoral degrees in the United States). From among the major aggregate branches of science, the United States led the Soviet Union in the number of specialists with advanced degrees in only the social sciences and the humanities. However, if agriculture and medicine are treated separately from the physical and life sciences, the United States showed an aggregate total for physics/mathematics, chemistry, environmental sciences, and biology about 20 percent higher than the Soviet Union.

• In 1950, the total number of natural scientists and engineers was about 20 percent greater in the United States than in the USSR; by 1974 (the latest year for which data are available for the USSR), however, the USSR had over twice as many natural scientists and engineers as the United States. This change in the relative number of scientists and engineers in the United States and the Soviet Union was largely the result of the enormous increase in the number of engineers in the USSR during this period. While in 1950, there were approximately the same number of engineers in the two countries, by 1974 the USSR had over three times as many engineers as did the United States. In the natural sciences, while there were more than twice as many scientists in the United States as in the Soviet Union in 1950, by 1974 this discrepancy had become somewhat less pronounced so that there were only about one and one-half times as many natural scientists in the United States as in the Soviet Union.

• The average annual rate of growth of full-time equivalent sci-entists and engineers engaged in R&D for the Soviet Union from 1961 to 1976 was about 8 percent, while for the United States it was only slightly over two percent. While estimates for 1960 indicate that the United States had more scientists and engineers engaged in R&D than did the Soviet Union, by 1976, depending on which estimate is used, the number of scientists and engineers engaged in R&D in the Soviet Union was from 1.3 to 2.3 times the number in the United States.

• Between 1950 and 1974, the average annual rate of growth in the number of natural scientists was slightly higher in the Soviet Union (6.8 percent) than in the United States (5.3 percent). The average annual rate of growth in the number of engineers, however, was more than twice as high in the Soviet Union (9.3 percent) as in the United States (4.3 percent). In the United States, the rate of growth of natural scientists exceeded the rate of growth of engineers, while in the Soviet Union, the reverse occurred.

• Since 1964, U.S. expenditures on R&D as a percentage of GNP have declined from 2.97 percent to 2.25 percent in 1978, a percentage drop of about 24 percent. By contrast, the Soviet Union appears to have been increasing the emphasis on performance of R&D in the economy. From 1963 to 1977, the United States showed an average annual decline in expenditures on R&D as a percentage of GNP of − 1.2 percent, while the Soviet Union showed an average annual rate of growth of 1.8 percent.

The above trends, which are primarily quantitative in nature, show dramatic increases in the number of scientists and engineers employed in the economy of the Soviet Union as well as in the number of students being trained in science and engineering fields at all stages of the educational process. Many of the comparative data on education and employment of scientists and engineers shown above, particularly those relating to breakdowns by field of science, do not extend beyond the mid-1970s. While more recent data are available for the United States, use of U.S. data was limited to those years for which comparable Soviet data are available. However, given the Soviet emphasis during the Tenth Five-Year Plan (1976-1980) on intensification of production with its objective of raising the qualifications and professional skills of workers, it can reasonably be assumed that, if anything, the Soviet Union has increased its gains in the number of scientific and engineering personnel relative to the United States during the last half decade.

C. Qualitative Considerations

As important, if not more so, however, are the qualitative aspects of the training and utilization of scientists and engineers. Unfortunately, the information available in the literature and that transmitted by the Soviet participants in the Science Policy Working Group exchanges did not permit an in-depth qualitative comparison of training in the two educational systems. This analysis should be high in the research priorities of American scholars. The structure of the Soviet higher-education program, however, does not appear to have undergone any major changes that would call into question the conclusion of earlier studies that the level of attainment of science and engineering graduates of full-time programs in Soviet higher educational establishments appeared to be about the same as, and occasionally higher than, the level of attainment of science and engineering graduates at the baccalaureate level in the United States. Approximately 40 percent of Soviet enrollment in higher educational establishments, however, is still in evening and correspondence programs, the quality of which is generally conceded to be substantially below that of full-time programs.

Other qualitative considerations that have emerged from the research that somewhat mitigate the impact of the dramatic quantitative differences in the growth of cadres of scientists and engineers in the Soviet Union relative to the United States are the following: (1) the extremely narrow specialization that characterizes the Soviet higher educational program; (2) the misutilization of many engineers, scientists, and other skilled R&D personnel in the Soviet economy; (3) the relative inflexibility of the Soviet economy with respect to the mobility of scientists and engineers; and (4) the deficiencies in Soviet methodologies and techniques for adequately forecasting requirements for scientists and engineers despite their necessity for the effectiveness of the Soviet planning process.

With respect to the first factor, the heavy Soviet concentration on narrow occupational training represents a marked difference from U.S. higher education. The extremely narrow specialization characteristic of Soviet higher education has often been faulted for its failure to provide scientists with the ability to master new knowledge, assimilate new research methods, and cope with technological change. In addition, such narrowly specialized training is highly susceptible to obsolescence and has for this reason been the subject of frequent controversy in the Soviet Union. While the high degree of specialization and applied functional orientation of the Soviet higher educational process may be an asset in the development of specialists with the ability to attain the short-term technological targets of the Soviet economic plan, the more flexible, theoretical, broader-based education received in U.S. higher educational institutions may produce specialists who are better prepared to meet the longer-term goals of a society with an ability to innovate, an adaptability to technological change, and a greater latitude for interfield mobility as the demands of the economy change.

Secondly, while the growth in scientific and engineering personnel in the Soviet Union has been quite dramatic, there is significant evidence from both Soviet and Western sources of misutilization of scientists, engineers, and other skilled R&D personnel in many fields of training. Highly trained personnel are known to spend a considerable portion of their time on administrative and support functions that could often be performed by less-qualified or auxiliary personnel. There is also evidence of mismatches between the needs of the economy for trained persons and the number and distribution by fields of the specialists the Soviet Union is producing. Inefficiencies result from training too many or too few specialists in particular fields; from having specialists, particularly engineers, perform jobs below their skill levels; and from low productivity of specialists because of the institutional framework in which they work. Poor incentives, bureaucracy, poor information collection and dissemination all hamper the effectiveness of Soviet specialists, even when they are appropriately trained and placed.

Turning to the third qualitative factor bearing on the quantitative comparisons of scientists and engineers in the United States and the USSR, it can be said that mobility of scientists and engineers in the two economies reflects many of the same causal factors and many of the same consequences; on the average, however, given the limited data available on Soviet mobility, scientists and engineers in the United States seem to be more fluid in the major types of mobility described. The opportunities for mobility seem to be greater in the United States. The Soviet system is more restrictive, the consequence of which is less effective utilization of scientific and engineering talent.

Finally, while methodologies for forecasting scientists and engineers in both the United States and the Soviet Union have major deficiencies, the U.S. methodologies are further developed to meet the requirements of the U.S. economy than the Soviet models are for their needs. In addition, the consequences of their absence or inaccuracies are less in the U.S. than the Soviet case. The market economy adjusts to changing conditions more rapidly than the planners can adjust their goals and force or induce their implementation. The Soviet forecasters have not, as yet, developed functional models appropriate to their economy and capable of being integrated into their planning process. Both characteristics are necessary to meet their needs.

In short, if a country's technological prowess were only dependent upon the number, scope, and sequence of science, mathematics, and engineering courses, then one might conclude that the Soviet Union is superior to the United States. But there are other factors, crucial ones, which must be considered before such a conclusion can be drawn. The Soviet Union has made remarkable strides in its educational system, particularly in the scientific and engineering fields, but there is evidence that suggests that the curriculum may be flawed both in implementation and in content. In the first case, the quality of instruction does not always match the quality of the curriculum. In the latter case, as stated above, the curriculum in higher and graduate education may be too narrow. This observation leads to a second: if the Russians are successful at training their scientists and engineers, they seem less successful at utilizing them efficiently.

It should be noted, however, that the quality of science, mathematics, and engineering education in the United States has also come under attack. A recent study prepared for the White House by the National Science Foundation and the Department of Education, entitled "Science and Engineering Education for the 1980s and Beyond," expresses the concerns of many American scholars about the current status of the teaching and learning disciplines which prepare students for employment in an increasingly scientific and technological world.

The White House study summarizes a number of problems in American science and engineering education. First, there is a serious shortage of secondary school teachers in mathematics and physical sciences. Because of the structure of U.S. education, there is no comprehensive science and mathematics curriculum required of all American school children. The drain of computer science and engineering graduates into the job market at the bachelor's degree level has resulted in a reduction of the supply of terminal-degree holders to teach and train future students in these disciplines. There is also evidence of a decreasing quality of baccalaureate and graduate programs due to overcrowding in classes, insufficient funding for laboratories and computer facilities, and relatively low salaries (vis-à-vis industry) to attract professors to teach in these programs. Finally, a decade-long slowdown in the growth of funding for research and development activities has begun to erode the ability of university science and engineering faculties to conduct innovative research.

D. Conclusion

While the above-noted problems in U.S. science and engineering education and utilization seem acute on the surface, they are even more
serious when contrasted with what is happening in the Soviet Union—
a nation in which political and economic institutions not only differ
markedly from those in the United States but also a nation which is
a strong economic and political competitor. The Soviet commitment
to the development of large cadres of highly trained scientists and
engineers is clear, as is the present broad strategy of the Soviet Union
to make use of such cadres to bring about quantitative improvements
in the economic sphere and thus increase the chance of Soviet success
in its competition with the Western world.

The United States does not face as dramatic a demographic constraint
as the Soviet Union in the next decade. There is a significant reservoir
of young people who, with appropriate incentives, could be redirected
into the scientific and engineering fields. A major concern raised by
the U.S. data is the decline in holders of doctorates in the engineering
fields who accept positions at colleges and universities as professors.
However, the most evident constraint seems to be the failure of the
leadership in the United States to grasp the necessity for a stated
national policy for development of scientific and engineering manpower and the support of that policy with a commitment of resources.
In the absence of any dramatic market shifts in the United States or
strong commitment by the federal government to provide incentives
for better-trained and more scientists and engineers, the United States
is not likely significantly to alter its comparative position with the
USSR in these critical manpower areas.

There is a need for concern in the United States, and that concern
should be manifested in a systematic and comprehensive review of our
training of Americans at all levels of education for an increasingly
scientific and technical world. Such a review will undoubtedly indicate
that changes are required, and we should be prepared to make them.
There will be a need for revisions of curricula; more financial resources
devoted to science and mathematics; more and better teachers at all
levels in the scientific and technical fields; more support for basic and
applied research; and, finally, the creation of an environment in which
all these can be accomplished with reasonable speed and coordination.
If anything less is accepted as a national objective, then there is cause
for alarm.

List of Specialty Groups and Specialties of USSR Higher Educational Institutions

INDEX NO.

1. *Geology and exploration of mineral deposits*

0101   Geology and prospecting of mineral deposits
0102   Geological surveying and searching for mineral deposits
0103   Geology and prospecting of oil and gas deposits
0105   Geophysical methods of searching and prospecting for mineral deposits
0106   Geochemistry (universities)
0107   Hydrogeology and engineering geology
0108   Technology and techniques of prospecting for mineral deposits

2. *Exploitation of mineral deposits*

0201   Mine surveying
0202   Technology and comprehensive mechanization of the underground exploitation of mineral deposits
0203   Technology and comprehensive mechanization of exploitation of peat deposits
0204   Enrichment of mineral deposits
0205   Technology and comprehensive mechanization of the exploitation of oil and gas deposits
0206   Construction of underground structures and pits
0207   Design and operation of gas and oil pipe-lines, gas-holders and oil reservoirs
0208   Construction of gas and oil pipe-lines, gas-holders and oil reservoirs
0209   Technology and comprehensive mechanization of open-cast working of mineral deposits
0210   Physical processes of mine production

0211	Drilling of oil and gas wells

3. *Power engineering*

0301	Electric power stations
0302	Electricity grids and networks
0303	Electricity supplies for industrial enterprises, cities, and agriculture
0304	Cybernetics of electricity grids
0305	Thermal electricity power stations
0306	Water processing and water-supply system of thermal and atomic power stations
0307	Hydro-power installations
0308	Industrial thermal power engineering
0309	Physics of thermal process
0310	Atomic power stations and installations
0314	High tension technology
0315	Electroenergy

4. *Metallurgy*

0401	Metallurgy of ferrous metals
0402	Metallurgy of non-ferrous metals
0403	Thermal technology and automation of metallurgical furnaces
0404	Casting of ferrous and non-ferrous metals
0405	Physical-chemical research techniques in metallurgical processes
0406	Physics of metals
0407	Metallography, equipment, and technology of thermal processing of metals
0408	Pressure-processing of metals
0411	Metallurgy and technology of welding

5. *Mechanical engineering and instrument construction*

0501	Technology of mechanical engineering, metal-cutting lathes, and tools
0502	Machinery and technology of casting processes
0503	Machinery and technology of pressure-processing of metals
0504	Equipment and technology of welding processes
0506	Mining machinery and outfits
0507	Peat-mining machinery and outfits
0508	Machinery and equipment for oil and gas fields
0509	Agricultural machinery
0510	Lifting, hoisting, and transportation machinery and equipment

0511	Construction and road-making machinery and equipment
0512	Wagon-building and wagon works
0513	Motor cars and tractors
0514	Shipbuilding and ship repairing
0515	Printing machinery
0516	Machinery and installations for chemical works
0517	Machinery and installations for food manufacture
0519	Machinery and mechanisms for forestry and woodworking industries
0520	Boiler construction
0521	Turbine construction
0522	Machinery and equipment for communication enterprises
0523	Internal combustion engines
0524	Ship engines and mechanisms
0525	Ship power installations
0526	Locomotive building
0527	Dynamics and durability of machinery
0528	Hydraulic machinery and automatic devices
0529	Refrigeration machinery, compressors, and installations
0530	Optical instruments and spectroscopy
0531	Precision instruments
0533	Cinematograph apparatus
0535	Aircraft construction
0537	Aircraft engines
0553	Hydro-aerodynamics
0558	Machinery and installations for cellulose pulp and paper production
0561	Chemical machinery apparatus building
0562	Mechanical equipment for firms producing building materials, products, and constructions
0563	Machinery and technology of the processing of products and parts from polymer materials
0566	Hydro-pneumatic automation and hydraulic drive
0567	Semi-conductor and electro-vacuum engineering
0568	Machinery and installations for the textile industry
0569	Machinery and installations for the light industry
0570	Machinery and installations for the production of chemical fibers
0571	Helicopter construction
0572	Mechanical equipment for factories of ferrous metallurgy
0573	Mechanical equipment for factories of non-ferrous metallurgy
0577	Machinery construction

6. *Electronics, electrical instrument construction, and automation*

0601	Electrical machinery and apparatus
0602	Electric traction and automation of traction-units
0603	Electric insulation and cable technology
0604	Semi-conductors and dielectrics
0606	Automation and telemechanics
0608	Mathematical and computing devices and appliances
0609	Gyroscopic instruments and appliances
0610	Electro-acoustics and ultra-sonic equipment
0611	Electronic instruments
0612	Industrial electronics
0613	Electro-thermal installations
0614	Lighting engineering and light sources
0615	Sound engineering
0617	Manufacture of aircraft instrumentation
0618	Electrical equipment for aircraft and motor tractors
0619	Electrical equipment of ships
0621	Technical operation of aircraft instruments and electrical equipment of aircraft
0627	Electronic medical instruments
0628	Electrical gearing and automation of industrial installations
0629	Semi-conductor instruments
0634	Electrification and automation of mining operations
0635	Automation and comprehensive mechanization of metallurgy
0636	Automation and comprehensive mechanization of machinery building
0638	Automation and comprehensive mechanization of building construction
0639	Automation and comprehensive mechanization of chemical technological processes
0640	Automation and mechanization of information processing and publishing processes
0641	Electronic physics
0642	Information measuring engineering
0643	Technology of special materials used in electronics
0645	Engineering electrophysics
0646	Automated control systems
0647	Applied mathematics
0648	Construction and production of computer apparatus
0649	Automation of thermal power processes
0650	Automation of the production and distribution of electrical power

7. *Radio engineering and communications*

0701	Radio engineering
0702	Automatic electrical communications
0703	Radio communications and broadcasting
0704	Radio-physics and electronics
0705	Construction and production of radio apparatus
0706	Technical operation of aircraft radio equipment
0708	Multi-channel electrical communications

8. *Chemical technology*

0801	Chemical technology of gas and oil processing
0802	Chemical technology of solid fuels
0803	Technology of inorganic substances and chemical fertilizers
0804	Chemical technology of rare and diffused elements
0805	Technology of electro-chemical processes
0806	Chemical technology of viscous materials
0807	Technology of basic organic and petrochemical syntheses
0808	Chemical technology of organic dyestuffs and intermediate products
0809	Chemical technology of biologically active compounds
0810	Chemical technology of plastics
0811	Chemical technology of varnishes, paints, and non-metallic coatings
0812	Technology of rubber
0813	Chemical technology of cinematographic-photographic materials
0819	Chemical technology of electronic and vacuum materials
0822	Chemical kinetics and combustion
0823	Technology of the separation and utilization of isotopes
0825	Radiation chemistry
0828	Technology of the processing of plastics
0829	Technology of chemicals used for plant protection
0830	Chemical technology of ceramics and refractory materials
0831	Chemical technology of glass and glass-works products
0832	Technology of electro-thermal products
0833	Technology of artificial fibers
0834	Basic processes in chemical manufacture and the cybernetics of chemistry
0835	Chemical technology of synthetic rubber
0836	Technology of recovery of secondary materials of industry

9. *Timber engineering and the technology of woodpulp, cellulose, and paper*

0901	Timber engineering

0902	Technology of wood-processing
0903	Chemical technology of woodpulp
0904	Chemical technology of cellulose and paper manufacture

10. *Technology of food products*

1001	Storage and technology of grain-processing
1002	Technology of bread-baking, macaroni and confectionery production
1003	Technology of saccharines
1004	Technology of fermentation processes
1005	Technology of wine-making
1006	Technology of fats
1007	Technology of canning and preserving
1008	Technology of sub-tropical cultures
1009	Technology of meat and meat products
1010	Technology of fish products
1011	Technology and organization of food services
1012	Fishing industry
1013	Ichthyology and fish-breeding
1015	Chemical technology of vitamins and albuminoid fermenting preparations
1016	Veterinary hygiene
1017	Technology of milk and dairy products

11. *Technology of consumer goods industries*

1101	Primary processing of fibrous materials
1102	Spinning of natural and artificial fibers
1103	Chemical technology and equipment of textile finishing mills
1104	Production of knitted goods
1105	Design and technology of ready-made garments
1106	Technology of leather and fur industry
1107	Technology of polymer film materials and synthetic leather
1108	Design and technology of leather products
1109	Technology of printing industry
1111	Weaving
1112	Construction of sewn goods
1113	Construction of leather goods
1114	Manufacture of non-woven textile materials

12. *Building construction*

1201	Architecture
1202	Industrial and civil construction
1203	

	Hydro-technical construction of river works and hydro-electric power stations
1204	Hydro-technical construction of waterways and harbors
1206	Urban construction
1207	Manufacture of building components and structures
1208	Gas and heat supply and ventilation
1209	Water supply and sewerage systems
1210	Construction of railways and roads, traffic engineering
1211	Motorways
1212	Bridges and tunnels
1213	Construction of airports
1217	Purification of natural water supplies and sewage disposal
1218	Technical utilization of buildings, equipment, and automatic systems

13. *Geodesy and cartography*
1301	Engineering geodesy
1302	Astro-geodsy
1303	Aerophotogeodesy (aerial photographic surveys)
1304	Cartography

14. *Hydrology and meteorology*
1401	Land hydrology
1402	Oceanography
1403	Hydrography
1404	Meteorology
1405	Agricultural meteorology

15. *Agriculture and forestry*
1501	Agricultural chemistry and soil science
1502	Agronomy
1503	Horticulture, market-gardening and viticulture
1504	Plant protection
1505	Sericulture
1506	Animal husbandry (zootechnics)
1507	Veterinary medicine
1508	Land utilization
1509	Mechanization of agriculture
1510	Electrification of agriculture
1511	Hydromelioration
1512	Forestry
1514	Mechanization of hydromelioration operations

16. *Transportation*
1601	Locomotives and their maintenance
1602	Electrification of railway transport

1603	Automatics, telemechanics, and signaling in railway transport
1604	Operation and maintenance of railways
1605	Municipal electric transportation
1606	Maritime navigation
1607	Inland waterway navigation
1608	Operation and maintenance of water transport
1609	Motor vehicle transport
1610	Operation and maintenance of aircraft and engines
1611	Operation and maintenance of air transport
1612	Operation and maintenance of ships' power installations
1613	Operation and maintenance of ships' electrical installations
1614	Mechanization of harbor-loading/unloading operations
1615	Industrial transport

17. *Economics*

1701	Planning of the national economy
1702	Planning of industry
1703	Economics and planning of supplies of technical materials
1704	Economics of labor
1705	Economics and organization of the mining industry
1706	Economics and organization of the oil and gas industries
1707	Economics and organization of the electric power industry
1708	Economics and organization of the metallurgical industry
1709	Economics and organization of the engineering industry
1711	Economics and organization of the chemical industry
1712	Economics and organization of the printing and publishing industry
1713	Economics of the cinematographic industry
1714	Economics and organization of the consumer goods industries
1715	Economics and organization of agriculture
1716	Planning in agriculture
1717	Economics and organization of procurement of agricultural produce
1718	Economics and organization of the foodstuffs industry
1719	Economics and organization of the timber industry and forestry
1720	Economics and organization of the woodworking and pulp and paper industry
1721	Economics and organization of building construction
1722	Economics and organization of municipal services
1723	Economics and organization of railway transport
1724	Economics and organization of water transport
1725	Economics and organization of motor vehicle transport

1726	Economics and organization of air transport
1727	Cinematics and the organization of book sales
1728	Economics and organization of communications
1729	Economics of trade
1731	International economic relations
1732	Merchandizing of industrial goods
1733	Merchandizing of food products
1734	Finance and banking
1736	Statistics
1737	Accounting
1738	Organization of the mechanized programming of economic information
1740	Accounting in agriculture
1741	Economics and organization of daily services
1742	Economics and organization of the radio-electronic industry
1743	Economics and organization of the construction materials industry

18. *Law*

1801	Jurisprudence
1802	International relations
1803	International law

19. *Public health and physical culture*

1901	Medical services
1902	Pediatrics
1903	Public health
1904	Stomatology
1905	Pharmacology
1906	Physical culture and athletics

20. *Specializations in universities*

2001	Russian language and literature
2002	Native languages and literatures
2003	Slavonic languages and literatures
2004	Romance and Germanic languages and literatures
2005	Oriental languages and literatures
2006	Classical philology
2007	Area studies of Eastern frontier countries
2008	History
2009	Historical and archival studies
2010	Political economy
2011	Philosophy
2012	Psychology

2013	Mathematics
2014	Mechanics
2015	Astronomy
2016	Physics
2017	Geophysics
2018	Chemistry
2019	Biology
2020	Zoology and botany
2022	Physiology
2024	Anthropology
2027	Journalism
2028	Literary composition
2029	History of the arts
2030	Geography
2033	Biophysics
2034	Biochemistry
2035	Economic cybernetics
2036	Structural and applied linguistics
2037	Document studies and the organization of administrative work in state establishments
2038	Scientific communism

21. *Specializations in pedagogical and cultural higher educational institutions*

2101	Russian language and literature
2102	Native languages and literatures of peoples of the USSR
2103	Foreign languages
2104	Mathematics
2105	Physics
2106	Biology
2107	Geography
2108	History
2109	Draftsmanship and drawing
2110	Pedagogy and psychology (pre-school)
2111	Defectology
2112	Cultural-education work
2113	Librianship and bibliography
2114	Physical education
2119	Music and singing
2120	General technical subjects and manual labor
2121	Pedagogy and teaching methods at primary level
2122	Chemistry

22. *Arts*

2201	Piano (organ)

2202	Orchestral instruments
2203	Fold instruments
2204	Singing
2205	Operatic and symphonic conducting
2206	Choral direction
2207	Composition
2208	Musical theory
2209	Dramatic theater and cinema acting
2210	Musical comedy acting
2211	Drama production, staging, and directing
2212	Musical theater production, staging, and directing
2213	Ballet production, staging, and directing
2214	Cinema production, staging, and directing
2215	Cinema operation techniques
2216	Theatrical techniques (staging effects) and staging of plays
2217	History and development of theater arts
2218	History and development of cinema arts
2219	Painting
2220	Drawing
2221	Sculpture
2222	Decorative and applied arts
2227	Art design and fashioning of products of the textile and light industries
2228	History and theory of figurative art
2229	Interior decoration and furnishing
2230	Industrial art
2231	Monumental and decorative art

Source: USSR Ministry of Higher and Secondary Specialized Education, *List of Specialties and Specializations of USSR Higher Educational Institutions,* Moscow, 1972.

Definitions and Concepts Underlying U.S. and Soviet Statistics on Financing Research and Development[1]

The borderlines between R&D activity as a whole and other closely associated or analogous activities, as well as between the various stages within the R&D cycle itself, have been very vague in common usage. Because of the subjectivity involved in the definition of the scope and internal divisions of the R&D process, the exact meaning of R&D statistics collected in different countries is not immediately obvious and may differ substantially from one country to another. Thus, an important issue in comparing data on the financing of R&D in the United States and the USSR is the clarification of similarities and differences in the concepts underlying the national statistics.

The National Science Foundation (NSF), as the federal agency responsible for collecting, analyzing, and summarizing data about the structure and growth of R&D expenditures in the United States, has formulated a complex set of definitions of various types of R&D activity for use in its detailed instructions to government agencies, firms, and other organizations for reporting purposes and in the structuring of data in its publications. The concept used in Soviet statistical reporting that most nearly approximates the concept of "total R&D" as generally used in the United States is that of "science" (*nauka*).

---

[1]A joint U.S.-Soviet effort to explore the comparability of Soviet statistics on expenditures on science with U.S. statistics on expenditures on R&D was one of the major objectives of the U.S.-USSR Subgroup on Financing Research and Development under the Joint Working Group in the Field of Science Policy. This Appendix is based on a study comparing the systems of financing R&D in the United States and the Soviet Union, "Financing Research and Development: U.S. and USSR," by Catherine P. Ailes, James G. Styles and Francis W. Rushing (SRI International, SSC-TN-4226-5, January 1981), which was prepared at the request of the U.S. members of the Joint Subgroup.

The Soviet "science" concept, however, differs from the U.S. concept of R&D in that it is wider in certain respects and narrower in others.

In addition to differences in Soviet practices compared to those of the United States in defining the research and development cycle as a whole, the distribution of R&D expenditures by type of R&D activity within the total cycle is not reported in Soviet national statistics, nor is such a distinction utilized for planning and financing purposes. While the division of the R&D process into stages is not utilized in the planning or reporting of R&D financing in the Soviet system, Soviet authors frequently refer to distinctions among types of R&D activity in their description and analysis of the R&D process, reflecting an increasing recognition of the desirability of such distinctions for planning and distribution of funds.[2] Although the conceptual division by stages is not described or labeled consistently by Soviet science policy analysts, nor does it exert any predictable influence on the recording of statistics, in general the breakdown is roughly along the same lines as the NSF definitions.

A. Basic Research

Basic research in the NSF classifications is described as research "directed toward increases of knowledge in science with the primary aim of the investigator being a fuller knowledge or understanding of the subject under study rather than a practical application thereof." When applied to the industrial sector, this general definition of basic research is modified by NSF to include specifically "original investigations for the advancement of scientific knowledge . . . which do not have specific commercial objectives, although they may be in the fields of present or potential interest to the reporting company." In the USSR, basic research (*fundamental' nye issledovaniya*) is similarly defined as "research that has a determinable practical importance but does not pursue specific practical goals."

B. Applied Research

Applied research is defined variously in NSF classifications as research "directed toward practical application of knowledge" or as "investigations directed to discovery of new scientific knowledge and which have specific commercial objectives with respect to either products or processes." Similarly, applied research (*prikladnye issledovaniya*) is

---

[2]Louvan E. Nolting, *Sources of Financing the Stages of the Research, Development, and Innovation Cycle in the U.S.S.R.*, Foreign Economic Report No. 3, U.S. Department of Commerce, (September 1973), p. 3.

generally defined in the USSR as "research that has a goal of obtaining economic or other practical effects."

Basic and applied research, as the two research stages of the R&D cycle, are comprehended under the general term "research" in NSF classifications and "scientific research work" (NIR: *nauchno-issledovatel'skaya rabota*) in the Soviet system.

C. Development

The NSF classifications define development as "the systematic use of scientific knowledge directed toward the production of useful materials, devices, systems or methods, including design and development of prototypes and processes." Development consists of technical activity related to non-routine problems encountered in translating research findings or other scientific knowledge into products or processes.

Similarly, in the Soviet Union, the development stage (*razrabotka*) of the R&D cycle is defined as "activity incorporating the results of applied research in (1) finished and tested technical drawings and other specifications, (2) prototypes of new machines, devices, and other products, and (3) new processes and methods of manufacture; and the testing of these results prior to their introduction into production or practical use."[3] Development in the Soviet Union is frequently referred to as "experimental-design development" or "experimental-design work" (OKR: *opytno-konstruktorskaya razrabotka* or *opytno-konstruktorskaya rabota*).

It is common in Soviet discussions to use the combination of scientific research and experimental-design work (NIOKR) as an analogue to the U.S. term R&D.

D. The Borderline Between R&D and Other Activities

Although the basic conceptual definitions of basic research, applied research, and development are roughly the same in the U.S. and Soviet systems, differences in U.S. and Soviet practices become apparent in determining the borderline between R&D activity and other closely associated or analogous activities.

The NSF concept of R&D ultimately is based on specific consideration of the purpose of the effort—in other words, whether or not an advance in knowledge or technology is intended. Thus, for example, although the development of new techniques for performing manage-

[3]Louvan E. Nolting, *The Financing of Research, Development and Innovation in the U.S.S.R., By Type of Performer*, Foreign Economic Report No. 9, U.S. Department of Commerce, (April 1976), p. 1.

ment functions, such as quality control or product testing, would be considered as R&D, the everyday use of such techniques would be regarded as production, marketing, or distribution. This distinction can be difficult to make because of the latitude for interpretation concerning the purpose or intent of the activity.

Other activities specifically excluded from R&D by the NSF classification include activities associated with the protection of property rights and the diffusion of new technology rather than with the actual innovation itself; adoption of new techniques and products developed elsewhere; collection of general purpose statistics, mapping and surveys, activities primarily concerned with the dissemination of scientific information; the training of scientific manpower; and the modification of existing technology, methods or processes that does not result in significant new knowledge or new approaches. When applied to the industrial sector, the NSF classification of R&D also excludes research in the social sciences, although social science R&D is included in NSF's classification of R&D performed by other sectors of the economy.

Because of the limited amount of published Soviet data on R&D expenditures and definitional uncertainties regarding the scope of available data, it is difficult to determine the precise points of divergence in the U.S. and Soviet practices of distinguishing R&D/"science" from other associated activities. However, it is reasonably well established that the scope of the Soviet "science" category is narrower than the NSF "R&D" category in some respects and broader in others. In particular, Soviet data on science expenditures exclude some of the more applied work at the final stages of development and testing, such as developing and testing of industrial prototypes and the development work performed by subdivisions of industrial enterprises, as described below:

- *Developing and Testing of Industrial Prototypes*
  One important respect in which the Soviet "science" category is narrower than the NSF R&D category is the development and testing of industrial prototypes (prototypes of machines or other devices undergoing evaluation and testing under production conditions). In the U.S. system, the boundary line for what is treated as development is the development of the prototype; once it is determined that a prototype is operative and the product or device can be placed in productive activity, the development phase is considered to be over and the production phase begins. The Soviet "science" category, however, is explicitly stated in Soviet

sources to exclude industrial prototype development and testing. This difference in reporting systems may be significant, for a substantial portion of R&D costs in the United States are associated with development and testing of prototypes.[4]

- *Development Work by Enterprise Units*
  Another example where Soviet treatment is narrower than the U.S. concept is the Soviet exclusion of R&D and technological improvement performed by the laboratories, design bureaus, and experimental facilities attached to industrial enterprises or plants. Such organizations appear to conduct a significant amount of development and testing of industrial prototypes, an exclusion already noted above.[5]

The areas in which the Soviet "science" concept appears to be wider than the U.S. concept of R&D include the dissemination of scientific information; the training of scientific personnel; the modification of existing technology, methods, and processes; routine economic forecasts, analyses, and business planning; and the social sciences. Much of the above is excluded from R&D in the U.S. reporting system, but is included to some degree or other in the Soviet "science" category, as indicated in the following:

- *Dissemination of Scientific and Technical Information*
  Activities associated with the dissemination of scientific and technical information are specifically excluded from U.S. reported R&D data. Soviet practices in this regard, however, appear to differ. A great deal of activity involved in the dissemination of scientific and technical information is carried out by the All-Union Institute of Scientific and Technical Information (VINITI), which may be considered a "scientific institution" whose expenditures are included in Soviet R&D totals. In addition, many of the branch institutes whose titles include the designation "Scientific Re-

---

[4]Nolting, *Sources of Financing*, pp. 1-2. There is some question, however, as to the degree of similarity or difference of Soviet and U.S. practices in this regard. In reviewing this draft, Nolting referred to remarks made by Ye. I. Valuyev, Head of the Department of Financing and Capital Investment of the USSR State Committee for Science and Technology, in a 1978 conference of the U.S.-USSR Joint Subgroup on Financing R&D, stressing that only "industrial prototypes," not prototypes produced in scientific research and design institutes, are excluded from science expenditures. As industrial prototypes represent the final stage of new product testing prior to mass production and therefore must be assembled and adjusted in the production plant, Nolting stated that he was "not so sure that the equivalent of 'industrial prototypes' in U.S. plants are classified under R&D costs."
[5]Nolting, *Sources of Financing*, p. 1, and Robert W. Campbell, *Reference Source on Soviet R&D Statistics, 1950-1978*, National Science Foundation, (1978), pp. 10-11.

search Institute,'' such as the All-Union Scientific Research Institute on the Organization, Administration and Economics of the Oil and Gas Industry (VNII OENG), are heavily involved in the dissemination rather than the creation of new scientific and technical knowledge.

- *Training of Scientific Personnel*
  The NSF classification of R&D specifically excludes activities associated with the training of scientific manpower, whereas the Soviet science expenditures concept includes the costs associated with graduate training conducted at scientific institutions. Such costs predominantly relate to stipends paid to graduate trainees, but also include some supervisory expenditures.

- *Modification of Existing Technology, Methods, or Processes*
  The U.S. concept of R&D specifically excludes activities associated with modification of existing technology or processes that does not result in significant new knowledge or new approaches. The Soviet concept makes no such exclusion, and it appears that when such work is conducted by organizations classified as ''scientific,'' costs associated with this work are included under ''science'' expenditures.

- *Routine Economic Forecasts, Analyses, and Business Planning*
  This may be a point of significant difference between U.S. and Soviet concepts of R&D. It appears that a considerable amount of this kind of work is done by Soviet R&D organizations for their supervising ministries, and hence is included in Soviet ''science'' totals. Thus, a great deal of the work associated with developing the software aspects of ''systems of automated planning and management'' (ASUP, in its Russian acronym) is done by organizations that go by the title of R&D organizations in the USSR and presumably is included in reported Soviet science totals, although it is excluded from the U.S. R&D concept.

- *Social Sciences*
  The Soviet ''science'' category includes research conducted in the social sciences. The NSF categorization of R&D specifically excludes social science R&D in industry, although it is included when performed by other sectors of the economy, i.e., government, universities and colleges, and other nonprofit institutions. The NSF exclusion of social science R&D in industry, however, has only minor effects on total R&D performed by industry.

● *R&D Support Personnel*

One other point of difference in U.S. and Soviet practices regarding exclusions from R&D statistics is with regard to Soviet inclusion of data on scientific support personnel employed in R&D organizations but not themselves directly engaged in R&D activity. A U.S. analyst suggests that "there must be hundreds of thousands of such people in the Soviet statistics on employment in science, with their wages included in data for expenditures on science."[6]

Many of the differences in coverage between U.S. and Soviet R&D statistics noted above can be attributed to the fact that in the USSR, R&D is generally conceptualized as the activities of those organizations that are officially defined as engaged in R&D, whereas in the United States, the focus is more explicitly on the activity itself. It is this, for example, that explains the exclusion of development work done in enterprises and the inclusion of information activities which happen to be conducted by an organization established in the R&D "sector" in Soviet R&D totals.

E. Total U.S. and Soviet Expenditures on R&D

National expenditures for performance of R&D as a percent of gross national product for the United States and the Soviet Union from 1962 to 1977 are shown in Chapter VII, Table VII-2. For 1977, U.S. expenditures on R&D amounted to about 2.3 percent of GNP, while Soviet R&D expenditures were estimated to be almost 3.5 percent of GNP. Estimates for the USSR were prepared for the National Science Foundation by Robert W. Campbell.[7]

In describing his methodology for estimates of Soviet R&D expenditures for comparison with U.S. data, Campbell notes that there is little choice in selecting an estimate for the Soviet Union other than the "science" series, which is generally treated as the analogue to U.S. expenditures on R&D. While Campbell suggests possible adjustments that might be made to improve the correspondence between the Soviet "science" series and U.S. R&D data, he states that "given that the figures are both over- and understated in a conceptual sense, along various dimensions, it is difficult to conclude how they are biased on balance."Campbell goes on to say that his "personal conclusion

---

[6]Campbell, op. cit., p. 13.
[7]The methodology is described in R. Campbell. *Reference Source on Soviet R&D Statistics 1950-1978.*

is that they are significantly inflated compared to U.S. data," an opinion that is consistent with the general consensus of most U.S. analysts of the Soviet economy.

One of the major goals of the U.S.-USSR exchange program on financing R&D was to gain a better understanding of the scope and composition of R&D expenditures in both the United States and the Soviet Union, so as to provide a basis for comparing the level of R&D effort in the two countries and for documenting differences in emphasis. Little was gained in this direction, however, because the Soviet experts did not provide additional detail on either the scope or composition of R&D expenditures relative to what was already known from other sources. The U.S. survey report, however, reveals that on the U.S. side we also lack the informational base necessary for a comparison of the level of effort of the two countries. For example, whereas the major Soviet data series includes some capital expenditures in R&D totals, data are not available on U.S. capital expenditures. In addition, because of a strong index number problem, a ruble comparison as well as a dollar comparison of expenditures in the two countries should be made. This would require, however, a breakdown of the U.S. total by class of expenditure (i.e., categories for which dollar–ruble ratios could be developed), and we do not currently have an adequate breakdown of outlay categories on the U.S. side to be able to do this.

A list of the reports and other documents prepared as part of the Joint Subgroup's activities is provided below. They are available from the National Technical Information Service (NTIS).

Survey Reports

- Ye. I. Valuyev, L.S. Glyazer, V.L. Groshev, V.I. Kushlin, M.R. Kokonina, G.A. Lakhtin, V.G. Lebedev, Yu. K.·Petrov, S.V. Pirogov, S.M. Ryumin, "Unique Characteristics of Financing Science in the USSR," Moscow (December 1976).

- Francis W. Dresch and Robert W. Campbell, "The System of Financing Research and Development in the United States," SRI International, SSC-TN-4226-1 (September 1977).

Responses to Questions

- "Answers to the Questions of the American Experts on the Draft of the Soviet Report 'The Financing of Science in the USSR,' " Moscow (November 1977).

- "Response to Questions Posed by the Union of Soviet Socialist Republics on the Report on 'The System of Financing Research and

Development in the United States,' '' SRI International (3 November 1977).

Conference Reports
- "Report on the Conference of the U.S.-U.S.S.R. Joint Subgroup on Financing Research and Development: March 8, 9 and 10, 1978," SRI International (May 1978).

Summaries and Translations of Materials Prepared by the Soviet Side
- "Unique Characteristics of Financing Science in the U.S.S.R.," (translation of Soviet report), U.S. Joint Publications Research Service, JPRS L/7289 (29 July 1977).

- "Summary of the Soviet Report on the Unique Characteristics of the Financing of Science in the U.S.S.R.," (based on the JPRS translation); SRI International (February 1978).

- "Edited Translation of Answers to the Questions of the American Experts on the Draft of the Soviet Report, 'The Financing of Science in the U.S.S.R.,' '' SRI International (May 1978).

Comparative Analyses
- Catherine P. Ailes, James G. Styles and Francis W. Rushing, "Financing Research and Development: U.S. and USSR", SRI International, SSC-TN-4226-5 (January 1981).

Bibliography

A.   Documents of the Joint Subgroup on the Training and Utilization of Scientific, Engineering and Technical Personnel.

U.S.S.R. REPORTS

Survey Reports

Zhiltsov, E.; Krutov, V.; and Ryabinin, A. "The Training and Utilization of Scientific and Engineering-Technical Personnel in the USSR: Part I." Draft Report. Moscow: Faculty of Economics, Economics of Education Laboratory, 1976.

Zhiltsov, E.; Krutov, V.; and Ryabinin, A. "The Training and Utilization of Scientific and Engineering-Technical Personnel in the USSR: Part I." Revised Final Report. Moscow: Faculty of Economics, Economics of Education Laboratory, November 1977.

Mikulinskiy, S.R.; Kugel', S. A.; Maslennikov, V. I.; Mirskii, E. I.; and Schelisch, P. B. "The Training and Utilization of Scientific and Engineering-Technical Personnel in the USSR: Part II." Moscow: December 1976.

Kugel', S. A., and Maslennikov, V. I. "The Mobility of Scientific Personnel: Supplement to the Survey Report on 'Scientific Personnel in the USSR'." Working Material. Moscow: July 1977.

Responses to Questions

"Answers to Questions of the American Experts on the First Part of the Soviet Report on 'Training and Utilization of Scientific and Engineering-Technical Personnel in the USSR". Moscow: 1977.

"Answers to Questions of the American Side on the Survey Report, 'Training and Utilization of Scientific and Engineering-Technical Personnel in the USSR: Part II'." Moscow: December 1977.

Kugel', S.; Mirskii, E.; and Schelisch, P. "Answers to Questions Posed by the American Subgroup on the Training and Utilization of Scientific and Technical Personnel Regarding the Working Material, 'Mobility of Scientific Personnel: Supplement to the Survey Report Scientific Personnel in the USSR'." Institute for the History of Science and Technology, U.S.S.R. Academy of Sciences, April 1978.

241

U.S. Reports

Survey Reports

Ailes, Catherine P.; Dresch, Francis W.; Ellis, Hazel T.; Lieberman, Anne R.; and Kincaid, Harry V. "Training and Utilization of Scientific and Engineering Technical Personnel." SRI International: SCC-TN-4226-2, February 1977.

Dresch, Francis W.; Ellis, Hazel T.; and Barries, Joel L. "Training and Utilization of Scientific and Engineering Technical Personnel: Chapters 10-14." SRI International: SCC-TN-4226-2, March 1977.

"Training and Utilization of Scientific and Engineering Technical Personnel: Appendices." SRI International: SCC-TN-4226-3, November 1976.

Responses to Questions

"Responses to Questions Posed by the Union of Soviet Socialist Republics on Part One of the Report on 'The Training and Utilization of Scientific and Engineering Technical Personnel'." SRI International: August 1, 1977.

"Responses to Questions Posed by the Union of Soviet Socialist Republics on Chapters 10-14 of the Report on the 'Training and Utilization of Scientific and Engineering Technical Personnel'." SRI International: November 3, 1977.

Conference Reports

Ailes, Catherine P. "Report on the Conference of the U.S.-U.S.S.R. Joint Subgroup on the Training and Utilization of Scientific, Engineering and Technical Personnel: March 8, 9 and 10, 1978." SRI International: May 1978.

Rushing, Francis W. "Report on the Moscow Conference of the U.S.-U.S.S.R. Joint Subgroup on the Training and Utilization of Scientific, Engineering and Technical Personnel: June 26, 27 and 28, 1978." SRI International: August 1978.

Summaries and Translations of Materials Prepared by the Soviet Side

Ailes, Catherine P.; Ellis, Hazel T.; and Rushing, Francis W. "Summary of Soviet Report on the Training and Utilization of Scientific, Engineering and Technical Personnel in the USSR." Final Report (incorporating summary report of July 1977 on Part I and Part II of

Soviet Report and summary report of December 1977 on Mobility Supplement to Soviet Report). SRI International: November 1979.

"Tables from the Soviet Report on the Training and Utilization of Scientific, Engineering, and Technical Personnel in the USSR." (Translation) SRI International: February 1978.

"Description and Unedited Translation of New Material Contained in the Revised Training Portion of the Soviet Report on 'The Training and Utilization of Scientific, Engineering and Technical Personnel in the USSR'." SRI International: January 1978.

"Translation of Soviet Response to Questions Posed by the U.S. Regarding the Training Portion of the Report on 'The Training and Utilization of Scientific, Engineering and Technical Personnel in the USSR'." SRI International: February 1978.

"Translation of Soviet Response to Questions Posed by the U.S. Regarding the Utilization Portions of the Report on 'The Training and Utilization of Scientific, Engineering and Technical Personnel in the USSR'." SRI International: February 1978.

"Translation of Soviet Responses to Questions Posed by the U.S. Regarding the Soviet Report on the Mobility of Scientific Personnel." SRI International: July 1978.

B.   Sources Cited in the Text

Afanasyev, V. G. *The Scientific and Technological Revolution—Its Impact on Management and Education.* Moscow: "Progress Publishers," 1975.

Ailes, Catherine P.; Cole, James E.; and Movit, Charles H. "Soviet Economic Problems and Technological Opportunities." *Comparative Strategy*, Vol. I, No. 4 (1979).

Brezharv, L. I., On the Basic Problem of the Economic Policy of the CPSU at the Contemporary Stage. Moscow: Politizdat, 1975.

Campbell, Robert W. *Reference Source on Soviet R&D Statistics, 1950-1978.* Washington, D.C.: National Science Foundation, 1978.

Carey, David W. "Developments in Soviet Education" in *Soviet Economic Prospects for the Seventies.* U.S. Congress, Joint Economic Committee, 1973.

Cocks, Paul M. *Science Policy: USA/USSR Volume II, Science Policy in the Soviet Union.* Washington, D.C.: U.S. Government Printing Office, 1980.

Conlin, Joseph S., Jr. *Soviet Professional, Scientific and Technical Manpower.* Washington, D.C.: Defense Intelligence Agency, 1973.

Corson, E.M. *An Analysis of the Five-Year Physics Program at Moscow State University.* Washington, D.C.: U.S. Department of Health, Education and Welfare, 1959.

DeWitt, Nicholas. "Current Status and Determinants of Science Education in Soviet Secondary Schools." Washington, D.C.: National Academy of Science, April 1980.

————. *Education and Professional Employment in the U.S.S.R.* Washington, D.C.: National Science Foundation, 1961.

Duzhenkov, V. I. "Problemy territorialnoy organizatsii nauchnoy deyatelnosti," *Problemy deyatelnosti uchenogo i nauchnykh kollektivov*, Moscow-Leningrad, 1977.

Engineers Joint Council. "Engineering Manpower in the Soviet Union." *Engineering Manpower Bulletin.* 1977.

Finn, Michael G. *Science and Engineering Technicians in the United States: Characteristics of a Redefined Population, 1972.* Oak Ridge: Manpower Research Programs, Oak Ridge Associated Universities, February 1978.

Folger, John K.; Astin, Helen S.; and Bayer, Alan E. *Human Resources and Higher Education.* New York: Russell Sage Foundation, 1970.

Freeman, Richard B. "A Cobweb Model of the Supply and Starting Salary for New Engineers." *Industrial and Labor Relations Review* 29 (January 1976).

————. "An Empirical Analysis of the Fixed Coefficient 'Manpower Requirements' Model, 1960-1970." *Journal of Human Resources*, Vol. 15, No. 2, Spring 1980.

————. *The Market for College-Trained Manpower: A Study in the Economics of Career Choice.* Cambridge: Harvard University Press, 1971.

————. *The Over-Educated American.* New York: Academic Press, 1976.

————. "Supply and Salary Adjustments to the Changing Science Manpower Market: Physics, 1948-1973." *American Economic Review* 19 (January 1976).

Grigor'yev, A. "Whom Shall We Train in the Higher Educational Institution?" *Leningradskaya Pravda* (April 15, 1977) pp. 2-3.

Gurin, V. *Sotsialistichesky trud.* Moscow (May 1973): 152-155.

Gvishiani, D. M.; Mikulinsky, S. R.; and Kugel', S. A., eds. *The Scientific Intelligentsia in the USSR.* Moscow: "Progress Publishers," 1976.

Helgeson, Stanley L.; Stake, Robert E.; Weiss, Iris R.; et al. *The Status of Pre-College Science, Mathematics, and Social Studies Educational Practices in U.S. Schools: An Overview and Summaries of Three Studies.* Washington, D.C.: National Science Foundation Directorate of Science Education, July 1978.

Heuer, Jill E. *Soviet Professional, Scientific, and Technical Manpower.* Washington, D.C.: Defense Intelligence Agency, May, 1979.

Korol, Alexander G. *Soviet Education for Science and Technology.* New York: John Wiley and Sons, 1957.

Lubrano, Linda L., and Berg, John K. "Scientists in the U.S.A. and U.S.S.R." *Survey* 23. Winter 1977-78.

National Academy of Science. *Field Mobility of Doctoral Scientists and Engineers*. Washington, D.C.: National Academy of Science, National Research Council, Commission on Human Resources, December 1975.

———. *The Invisible University, Postdoctoral Education in the United States*. Washington, D.C.: National Research Council, National Academy of Science, 1969.

———. *Mobility of Ph.D.s: Before and After the Doctorate with Associated Economic and Educational Characteristics of States*. Research Division of the Office of Scientific Personnel, National Academy of Sciences, 1971.

National Center for Education Statistics. *Projections of Education Statistics to 1988-89*. April 1980.

National Economy of the USSR. Moscow: "Statistika." Annual Series.

National Education, Science and Culture in the USSR: 1971. Moscow: 'Statistika," 1972.

National Education, Science and Culture in the USSR: 1977. Moscow: "Statistika," 1978.

National Science Board. *Science Indicators 1978*. Washington, D.C.: National Science Foundation, 1979.

National Science Foundation. *Graduate Science Education: Student Support and Postdoctorals*. Washington, D.C.: National Science Foundation. (Annual Series.)

———. *Graduate Student Support and Manpower Resources in Graduate Science Education*. Washington, D.C.: National Science Foundation. (Annual Series.)

———. *National Patterns of R&D Resources, 1953-1977*. Washington, D.C.: National Science Foundation, April 1977.

———. *National Patterns of R&D Resources 1953-1978-79*. Washington, D.C.: National Science Foundation, 1978.

———. *National Patterns of Science and Technology Resources, 1980*, Washington, D.C.: National Science Foundation, March 1980.

———. *The 1972 Scientists and Engineers Population Redefined*. 2 Vols. Washington, D.C.: Government Printing Office, 1975.

———. *Science and Engineering Education for the 1980s and Beyond*. Washington, D.C.: National Science Foundation and Department of Education, October 1980.

———. "The Status of Pre-College Science, Mathematics and Social Studies Educational Practices in U.S. Schools." Washington, D.C.: National Science Foundation, 1978.

New York Times. "Foreigners Snap Up the High-Tech Jobs." July 5, 1981.

Nolting, Louvan E. "The Financing of Research Development and

Innovation in the U.S.S.R. by Type of Performer." *Foreign Economic Report No. 9*. Washington, D.C.: U.S. Department of Commerce, 1976.

————, and Feshbach, Murray. "R&D Employment in the USSR—Definitions, Statistics and Comparisons." *Soviet Economy in a Time of Change*, Vol. I. U.S. Congress, Joint Economic Committee, October 1979.

Nozhko, K.; Monozon, E.; Zhamin, V; and Severfsev, V. *Educational Planning in the USSR*. UNESCO: International Institute for Educational Planning, 1968.

Panferova, N. "Too Many or Too Few." *Literaturnaya gazeta*, November 23, 1977.

Price, Derek J. de Solla. "Measuring the Size of Science." Proceedings of the Israel Academy of Sciences and Humanities, Vol. 9, No. 6, 1969.

Remennikov, B. M. *The Higher School in the System of Reproduction of the Labor Force in the USSR*. Moscow, 1973.

Rushing, Francis W., and Ailes, Catherine. "An Assessment of USSR-US Scientific and Technical Exchange Programs." *Soviet Economy in a Time of Change*, Vol. II, U.S. Congress, Joint Economic Committee, October 1979.

Shapero, A., and Howell, R. P. "The Structure and Dynamics of Research and Exploratory Development in the Defense R&D Industry." Stanford Research Institute, 1966.

Soviet Economic Prospects for the Seventies. U.S. Congress, Joint Economic Committee, 1973.

Talley, Roger K. *Soviet Professional Scientific and Technical Manpower*. Washington, D.C.: Defense Intelligence Agency, 1976.

U.S. Bureau of the Census, Foreign Demographic Analysis Division. USSR population estimates for July 1 of each year (March 1977).

U.S. Department of Commerce. *Statistical Abstract of the United States*. Washington, D.C.: Government Printing Office. Annual Series.

U.S. Department of Health, Education and Welfare. *Digest of Education Statistics*. Washington, D.C.: National Center for Education Statistics. Annual Series.

————. *Earned Degrees Conferred*, Annual Series. Washington, D.C.: National Center for Education Statistics, 1978.

U.S. Department of Labor. *Employment of Scientists and Engineers, 1950-1970*. Washington, D.C.: Bureau of Labor Statistics, 1973.

————. *Occupational Outlook Handbook, 1976-77*. Washington, D.C.: Bureau of Labor Statistics, 1976.

————. *Occupational Projections and Training Data, 1980 Edition*. Washington, D.C.: Bureau of Labor Statistics, September 1980.

————. *A Survey of Scientific and Technical Personnel in Industry in 1969*. Washington, D.C.: Bureau of Labor Statistics, 1971.

————. "Technician Manpower: Requirements, Resources, and Training Needs." Washington, D.C.: Bureau of Labor Statistics, 1966.

The U.S.S.R. in Numbers in 1975. Moscow: "Statistika," 1976.

USSR: Trends and Prospects in Educational Attainment, 1959-1985. Washington, D.C.: National Foreign Assessment Center, June 1979.

Wolfle, Dael. *The Use of Talent.* Princeton: Princeton University Press, 1971.

Zaccaria, Joseph. *Theories of Occupational Choice and Vocational Development.* Boston: Houghton Mifflin Co., 1970.

Zaleski, E., et al. *Science Policy in the USSR.* Organization for Economic Co-operation and Development, 1970.

Zhiltsov, E., and others. "The Training and Utilization of Scientific, Engineering and Technical Personnel in the USSR, Part I." Moscow: Faculty of Economics, Economics of Education Laboratory, November 1977.

C. Other Reference Sources

Anodin, G. S. *The Determination of Specialist Requirements in Industry: The Case of the Coal Industry.* Moscow: Gosplanizdat, 1958 (in Russian).

Bermant, M. A.; Semenov, L. K.; and Sulitskiy, V. N. *Mathematical Models and the Planning of Education.* Moscow: Nauka, 1971 (in Russian).

Braginskiy, V. I., et al. *The Planning of Specialists Requirements of the National Economy.* Moscow: Gosplanizdat, 1959 (in Russian).

Gvishiani, D. M.; Mikulinskiy, S. R.; and Kugel, S. A., eds. *The Scientific-Technical Revolution and Structural Change of Scientific Cadres in the USSR.* Moscow: Nauka Press, 1973 (available in translation).

Helgeson, Stanley L.; Blosser, Patricia E.; and Howe, Robert W. "The Status of Pre-College Science, Mathematics, and Social Science Education: 1955-1975, Vol. I: Science Education." Columbus: The Ohio State University, Center for Science and Mathematics Education. (Mimeographed.)

Khokhlov, R. V., et al. *The Scientific-Technical Revolution and the Development of Higher Education.* Moscow: MGU Press, 1974 (in Russian).

Komarov, V. Ye. *The Economic Bases of the Training of the Specialists of the National Economy.* Moscow: Akademia Nauk, 1959 (in Russian).

————. *Economic Problems of the Training and Utilization of Specialist Manpower.* Moscow: Ekonomika, 1972 (in Russian).

————. *Economic Problems of the Training and Utilization of Specialist*

Manpower in the National Economy. Doctoral Dissertation, Moscow: Institute of Economics, Academy of Sciences, 1967 (in Russian).

———. "On the Rational Utilization of Specialist Manpower," *Voprosy Ekonomiki*, 9 (1966), pp. 15-25 (in Russian).

Kostin, L. A. *The Planning of Labor in Industry.* Moscow: Ekonomika, 1967 (in Russian).

Kugel', S. A.; Levin, B. D.; and Meleschenenko, Yu. S.. *Scientific Cadres of Leningrad.* Leningrad: Nauka Press, 1973 (in Russian).

Letenko, V.A.; Bryanskiy, G.A.; and Kuzyakov, N. V. *Directions for Diploma Projects.* Moscow: Higher School Press, 1976 (in Russian).

———. *Analysis of the Dynamics of Growth, Forecasts of Training, and the Structure of Scientific-Pedagogical Cadres at the Tula Polytechnic Institute.* Moscow: Minvuz, 1977 (in Russian).

Levin, B. D., and Perfil'yev, M. N. *Cadres of the Administrative Apparat in the USSR.* Leningrad: Nauka Press, 1970 (in Russian).

Mamyedov, Ya. "Questions on the Fluctuation of Engineering-Technical Workers and Ways of Its Reduction (Using the Example of the 'Turkmenneft' Industrial Association)." *Izvestiya Adademii Nauk Turkmenskoy SSR Seriya: Obshchestvennykh Nauk*, 1 (1973), pp. 65-67 (in Russian).

NIITruda. *The Construction of a Management Apparatus in Enterprises and in Industrial Associations: Interbranch Methodological Recommendations.* Moscow: NIITrude, 1974 (in Russian).

Noah, Harold J. *Financing Soviet Schools.* New York: Columbia University, 1966.

Savichev, K. P. *The Training and Distribution of Young Specialists in the USSR.* Moscow: Vyshaya Shkola, 1972 (in Russian).

Scientific Research Institute for the Problems of Higher Education. *Branch Methodologies for Determining Requirements for Specialists.* Moscow: Minvuz, 1977 (in Russian).

———. *Methodology for Forecasting the Number of Specialists with the Aid of the Exponential Smoothing Method.* Moscow: Minvuz, 1977 (in Russian).

———. *Normative Research Methods of Planning the Training of Specialists Cadres.* Moscow: Minvuz, 1976 (in Russian).

———. *Several Indicators of the Effectiveness of Scientific Activity in Moscow VUZy.* Moscow: Minvuz, 1977 (in Russian).

———. *Several Methods for Calculating the Expenditure on the Training of One Specialist in the VUZ.* Moscow: Minvuz, 1977 (in Russian).

Shishkina, L. I. *Social-Legal Questions of the Professional Orientation of Youth.* Leningrad: Leningrad University Press, 1976 (in Russian).

Slezinger, G. Ye. *Labor in the Management of Industrial Production.* Moscow: Ekonomika, 1967 (in Russian).

Sochi Conference Reports, Sochi, U.S.S.R.

Agursky, Michael. *The Research Institute of Machine-Building Technology.* Hebrew University of Jerusalem, Jerusalem, 1976 (in Russian).

Davidov, L. D. *Five Year Planning of Training and Increasing the Managerial Cadre.* Sochi Conference Paper, 1976.

Korolev, Y. A., and Kazakov, V. N. *Planning the Branch System of Increasing the Qualification of Managers.* Sochi Conference Materials: 1976.

Lilyzyev, V. C. "Economic-Mathematical Model of the Determination of Prospective Requirements for Specialists." Paper presented at 1976 Sochi Conference: October, 1976 (in Russian). (Xerox.)

Shuruyev, A. S. "Planning the Training of Specialists in the USSR." Paper presented at 1976 Sochi Conference: October, 1976 (in Russian). (Xerox.)

Solodkov, M. V. "Problems of Planning the Training of Economists in the USSR." Paper presented at 1976 Sochi Conference: October, 1976 (in Russian). (Xerox.)

Zhil'tsov, E. N., and Chuprunov, D. I. "Setting Norms for Engineering-Administrative Labor as a Basis for Planning the Preparation of Specialists with Higher Education." Sochi Conference Paper, 1976.

Stake, Robert E.; Easley, Jack; et al. "Case Studies in Science Education, Vol. I: The Case Reports, Vol. II: Design, Overview and General Findings." Urbana: University of Illinois, Center for Instructional Research and Curriculum Evaluation. (Mimeographed.)

Suydam, Marilyn N., and Osborne, Alan. "The Status of Pre-College Science, Mathematics, and Social Science Education: 1955-1975, Vol. II: Mathematics Education." Columbus: The Ohio State University, Center for Science and Mathematics Education. (Mimeographed.)

Tashchev, A. K., ed. *Engineering Labor in the Socialist Society.* Moscow: Mysl Press, 1976 (in Russian).

Topchiyev, A. V. *Scientific Cadres in the USSR.* Moscow: Academy of Sciences Press, 1959 (in Russian).

Volkov, F. M.: Nemchenko, V. S.; and Yagodkin, V. N., eds. *The Production of Labor Force and Increasing the Efficacy of the Utilization of Labor Resources.* Moscow: M. G. U. Press, 1971. (in Russian).

Weiss, Iris R. "Report of the 1977 National Survey of Science, Mathematics, and Social Studies Education." Research Triangle Institute. (Mimeographed.)

Index